AF782536

CONTENTS

LAW ENFORCEMENT ACCESS TO COMMUNICATION SYSTEMS IN THE DIGITAL AGE

WEDNESDAY, SEPTEMBER 8, 2004

HOUSE OF REPRESENTATIVES,
COMMITTEE ON ENERGY AND COMMERCE,
SUBCOMMITTEE ON TELECOMMUNICATIONS
AND THE INTERNET,
Washington, DC.

The subcommittee met, pursuant to notice, at 10 a.m., in room 2322, Rayburn House Office Building, Hon. Fred Upton (chairman) presiding.

Members present: Representatives Upton, Stearns, Cox, Shimkus, Pickering, Buyer, Walden, Terry, Barton (ex officio), Wynn, McCarthy, Rush, and Stupak.

Staff present: Howard Waltzman, majority counsel; Will Norwind, majority policy coordinator; William Carty, legislative clerk; and Peter Filon, minority counsel.

Mr. UPTON. Good morning. We have a lot of hearings today, and a number of our colleagues are allegedly on the way, and I will just make a motion for unanimous consent that all members' statements on the subcommittee will be put into the record in their entirety, if by some chance they do not get here by the time our opening statements are concluded.

Today's hearing is entitled "Law Enforcement Access to Communication Systems in a Digital Age." I want to put today's hearing into context.

In order to realize their full potential, I believe that broadband and VoIP services should be free from unwarranted economic regulation. However, there are other types of regulation which promote important policy objectives. For example, today we will be examining law enforcement access to communication systems, which is facilitated through CALEA.

More specifically, we will examine whether and how such access should be provided in the digital age. Unfortunately when it comes to telecommunications technology, many terrorists are not as primitive as their evil and demented world view.

In fact, law enforcement raises the specter of terrorists exploiting perceived technological gaps with respect to certain services for which telecommunications carriers are unable to provide or are unable to provide in a usable form the content of communications or related information as required by a court order.

According to law enforcement, certain services will become the preferred means of communications for terrorists precisely because of such perceived technological gaps in the ability of law enforcement to get access.

In our post 9/11 world, we ignore this specter at our own peril. That is why we must insure that law enforcement has adequate access to digital communications like broadband and VoIP.

Having said that, we must work carefully with telecommunications carriers and manufacturers to insure that the technological standards for providing such access are driven by industry which is in a better position than the government to find workable ways to build the proverbial mousetrap without stifling innovation in this relatively nascent and dynamic marketplace.

While there may be an inherent and understandable tension between law enforcement and industry, this has to be a partnership if we are to put our best foot forward in the name of Homeland Security.

I look forward to hearing from today's witnesses, and I want to thank them in advance for their testimony which we received last night.

And I will recognize for an opening statement my colleague from Oregon, Mr. Walden.

Mr. WALDEN. Thank you, Mr. Chairman, for holding this hearing. I am going to waive an opening statement in lieu of time for questions.

Mr. UPTON. Mr. Buyer.

Mr. BUYER. Reserve my time.

Mr. UPTON. Mr. Shimkus.

Mr. SHIMKUS. The same.

Mr. UPTON. Okay. Well, that is terrific.

[Additional statements submitted for the record follow:]

PREPARED STATEMENT OF HON. PAUL E. GILLMOR, A REPRESENTATIVE IN CONGRESS FROM THE STATE OF OHIO

Thank you Mr. Chairman, for yet another opportunity to lay the groundwork for addressing the insurgence of new technologies under current telecommunications rules.

Today, where we will learn more about the critical needs of law enforcement agencies to easily access information from emerging technologies such as Voice over Internet Protocol (VoIP) and broadband services for crime-fighting purposes, we must also keep in mind the importance of further developing these new modes communication in order to continue to create more competition and meet customers' demands.

I welcome the well-balanced panel of witnesses and again, look forward to hearing about how our panel can contribute to striking this delicate balance, one of protecting our homeland and spurring technological innovation.

Again, I thank the Chairman and yield back the remainder of my time.

PREPARED STATEMENT OF STATEMENT OF HON. BARBARA CUBIN, A REPRESENTATIVE IN CONGRESS FROM THE STATE OF WYOMING

Thank you, Mr. Chairman.

Just a few years ago Voice over Internet Protocol (VoIP) seemed to be a technology for a future telecommunications generation. But it's here now, and is making us all wrestle with where VoIP fits in the framework of our communications infrastructure. On one hand, it provides the promise of true competition that promises to benefit consumers across America. On the other, it calls into question a host of legacy regulations that make up the telecom landscape today.

Part of the VoIP debate is where social obligation regulations fit into the picture. One that is of particular concern, and is important to the safety of our homeland, is the Communications Assistance for Law Enforcement Act (CALEA). This act gives our law enforcement organizations tools they need to catch the bad guys. It used to be that these bad guys were just your garden variety criminals. Now they are terrorists who work to plot their next attack on America with whispered voices in shadowy places around the world—and here at home.

That's why it is important to me that we not allow a means of communication that would skirt the law, and invite terrorists to communicate with impunity. I am interested to hear from our witnesses on this matter and to determine the proper role for this Committee and this Congress regarding VoIP and CALEA.

Thank you again, Mr. Chairman for opening this dialog and look forward to hearing from our witnesses. I yield back the balance of my time.

PREPARED STATEMENT OF HON. JOE BARTON, CHAIRMAN, COMMITTEE ON ENERGY AND COMMERCE

Mr. Chairman, thank you for calling this hearing today. As this committee examines the rules that should apply to broadband services and networks, it is critical that we understand the implications of any changes in the rules for law enforcement access to communications systems.

There are three primary goals we need to keep in mind in approaching this issue. First, we must not permit broadband or VoIP services to become the communications-medium-of-choice for terrorists because of the absence of electronic surveillance capabilities for law enforcement. Second, however, we must not stifle new technologies by burdening them with unachievable rules. And, third, we must protect consumer privacy. While lawfully executed court orders must enable law enforcement to have access to certain call content and call-identifying information, this must not lead to a wanton invasion of consumer privacy.

I believe that these three goals are achievable and do not have to be mutually exclusive. And I believe that the FCC has started down a path that will enable us to achieve these goals.

Lawfully conducted electronic surveillance is a critical component of effective law enforcement. New, broadband technologies have tremendous promise for our society. We need to make sure that these technologies are used for the benefit of society and are not used by terrorists to evade detection.

Mr. Chairman, thank you for holding this hearing. I look forward to the testimony of our witnesses.

PREPARED STATEMENT OF HON. JOHN D. DINGELL, A REPRESENTATIVE IN CONGRESS FROM THE STATE OF MICHIGAN

Mr. Chairman, thank you for holding this hearing today on the Communications Assistance for Law Enforcement Act, known as CALEA. CALEA is a critical tool for law enforcement agencies in maintaining the access communications of terrorists, drug traffickers, organized crime syndicates, and other criminals. CALEA is vital to our nation's security.

Although CALEA was written ten years ago in a mostly analog world, Congress understood that new digital communications technologies were on the horizon. Accordingly, CALEA was written with sufficient flexibility to preserve the government's ability to access many communications among users of advanced digital networks.

Both the world and technology have changed significantly since 1994. The spread of terrorism has in many respects made the world a much more dangerous place. Moreover, new technologies have spread—new digital broadband networks have come on line and new digital applications, such as Voice over Internet Protocol telephone service, are riding over these networks.

While providing tremendous opportunities for consumers, these technologies may unintentionally provide terrorists, drug traffickers, and other criminals new ways to evade detection by law enforcement. Not only are criminals adept at exploiting new technologies for illegal purposes, but the uncertainty surrounding CALEA's application to new technologies is only exacerbating the situation. In fact, it has gotten to the point where Deputy Assistant Attorney General John G. Malcolm was quoted earlier this year in a New York Times article as saying that he was "aware of instances in which law enforcement authorities have not been able to execute intercept orders because of this uncertainty." It is imperative that the Bush Administration and the Federal Communications Commission (FCC) fully implement CALEA.

Under CALEA, law enforcement agencies have the authority to gain access to communications information being transmitted by telecommunications carriers, the definition of which is much broader than the definition of such carriers in the Communications Act. Moreover, CALEA provides the Commission authority to bring within the scope of CALEA new services that act as a replacement for a substantial portion of local exchange service. The CALEA statute is clear.

Last month the FCC finally issued a notice of proposed rulemaking that made several tentative conclusions, including that both facilities-based providers of broadband service and "managed" VoIP services are subject to CALEA. I would note, however, that it has been ten years since CALEA became law and three years since terrorists attacked the World Trade Center and Pentagon. Lives are at stake. Why has it taken the Commission so long to act on such an important issue?

Similarly, since September 11th, neither the Commission nor the Bush Administration has developed a comprehensive nationwide plan to ensure the reliability, redundancy and interoperability for communications systems, especially those of public safety.

Despite the delay in issuing its proposed rule, I am pleased that the Commission has been mindful of Congress' three underlying goals in CALEA: First, that government maintains the ability to intercept communications involving new technologies; second, that the privacy of individuals is protected; and third, that unnecessary burdens on the development of new technologies and services are avoided. In this dangerous new world, it is important that undue delays also be avoided in the implementation of CALEA so that the government maintains the ability to protect its people from those who seek to do them harm.

Mr. UPTON. We are joined by a very distinguished panel. We are actually going to be starting with Ms. Laura Parsky, Deputy Assistant Attorney General for Criminal Division, U.S. Department of Justice; followed by Mr. Marcus Thomas, Deputy Assistant Director of the Federal Bureau of Investigation; Mr. Julius Knapp, Deputy Chief of the Office of Engineering and Technology, FCC; Mr. Stewart Baker, Steptoe & Johnson; Dr. Richard Green, President and Chief Executive Officer of Cable Television Labs; and Mr. James Dempsey, Executive Director of the Center for Democracy and Technology.

Ladies and gentlemen, your testimony will be made part of the record in its entirety. We would like you to summarize it if you can in the time of about 5 minutes on the clock.

We will start with you, Ms. Parsky. Thank you for being with us today.

STATEMENTS OF LAURA H. PARSKY, DEPUTY ASSISTANT ATTORNEY GENERAL, CRIMINAL DIVISION, U.S. DEPARTMENT OF JUSTICE; MARCUS C. THOMAS, DEPUTY ASSISTANT DIRECTOR, FEDERAL BUREAU OF INVESTIGATION; JULIUS P. KNAPP, DEPUTY CHIEF, OFFICE OF ENGINEERING AND TECHNOLOGY, FEDERAL COMMUNICATIONS COMMISSION; STEWART A. BAKER, STEPTOE & JOHNSON; RICHARD R. GREEN, PRESIDENT AND CHIEF EXECUTIVE OFFICER, CABLE TELEVISION LABORATORIES, INC.; AND JAMES X. DEMPSEY, EXECUTIVE DIRECTOR, CENTER FOR DEMOCRACY AND TECHNOLOGY

Ms. PARSKY. Good morning Chairman Upton and members of the subcommittee. I appreciate the opportunity to speak with you today as you are considering the appropriate regulatory framework for new communications technologies. These advanced technologies, including high-speed broadband Internet access and telephone service that uses Voice over Internet Protocol, or VoIP, promise to contribute to increased American productivity and to offer consumers the convenience of reasonably priced, high-quality service.

One report indicates that a majority of U.S. households now use some means of high-speed Internet access. In addition, Internet telephony is attracting more and more customers every day.

The administration fully supports the rapid and widespread deployment of such new services. We also welcome and applaud your efforts and the efforts of others in Congress as you carefully debate the proper regulatory environment for them. To automatically apply old-fashioned and likely outdated principles to a new way of doing business is sure to hamper the development of these promising and potent technologies.

However, in devising new principles for governing these new technologies, we must preserve those safeguards that are critical to our national security and public safety.

The core issue here is responsibility: responsible government and responsible citizenship. By reevaluating traditional regulation of communications systems, the government is acting responsibly. Likewise, those who develop and provide such communications services must also assume responsibility.

The Communications Assistance for Law Enforcement Act, CALEA, was drafted 10 years ago when Congress could not have anticipated the details of today's communications revolution. However, Congress did have the foresight to predict that such a communications revolution would take place.

CALEA requires that, as new technologies are developed, providers act responsibly by engineering their systems in a way that allows law enforcement to execute court-ordered electronic surveillance. As communications technology has progressed, some carriers have never questioned their legal obligations under CALEA or their corporate obligations to act responsibly where public safety and national security are at risk. For each and every carrier in this category, we recognize and applaud their leadership and responsibility.

Unfortunately, however, there are also some carriers who have deployed their technologies without regard to law enforcement's ability to execute court-ordered electronic surveillance and without regard to their corporate responsibility where public safety and national security are at risk.

Because of the existence of carriers in this latter category, we have been forced to petition the Federal Communications Commission to affirm the legal obligations of carriers to comply with CALEA, and for the Federal Communications Commission to meet noncompliance with robust enforcement actions.

CALEA's obligations are even more important today than they were when the statute was enacted 10 years ago. While many carriers act responsibly and in the public interest without the need for compulsory process, there will always be some businesses that will choose to operate without regard to such concerns. Because savvy criminals and terrorists seek out those businesses, we must take steps to eliminate the vulnerability in our national security and public safety created by those businesses.

CALEA and the robust enforcement of CALEA will help accomplish this critical goal. It is important to recognize that CALEA itself does not authorize any electronic surveillance. When enacting

CALEA, Congress recognized that law enforcement has had the authority to conduct wire taps pursuant to a court order since 1968.

Well prior to CALEA, the authority extended to intercepts of voice, data, fax, E-mail and any other form of electronic communications, and Congress expressly stated that CALEA would not expand that authority.

What CALEA does is to help ensure that as new communications technologies are developed, carriers using those technologies are capable of isolating and providing to the government, in real time, communications and related information as required by court orders. Electronic surveillance itself is a law enforcement tool of last resort and, to use it, Federal and State governments must meet numerous constitutional, statutory, and regulatory requirements.

Electronic surveillance is, however, usually critical to those investigations where it is used successfully. Such wiretaps are used to identify participants in organized crime and obtain evidence about their specific criminal activities, to identify participants in major drug offenses and seize significant quantities of contraband drugs and currency, to solve or prevent murder and other violent crimes attendant to organized crime and drug trafficking, to solve or prevent crimes involving the sexual exploitation of children, and, increasingly, to solve or prevent terrorist offenses.

In a recent child sexual exploitation investigation in Oklahoma, for example, investigators obtained judicial authorization to intercept all wire communications of a pimp who traveled interstate in order to sell children for sexual activity. The pimp was recorded talking about grooming children to become prostitutes, physically beating his victims into compliance, and marketing the children as prostitutes in numerous States.

Further, the electronic surveillance helped identify a national child prostitution network and generated investigations in other States. To date, the U.S. Attorney's Office in Oklahoma City has federally charged nine defendants for sexually exploiting children and more indictments are pending. Significant State charges have also been filed against ten perpetrators of these horrible crimes. Already three children, one from Las Vegas, one from New Mexico, and one from Oklahoma, have been rescued by law enforcement thanks to the electronic surveillance.

Moreover, probably thousands of physical and sexual assaults upon children have been prevented as a result of these prosecutions that were dependent on electronic surveillance.

Electronic surveillance is also critical to the Department's highest priority, fighting the war on terrorism. The cell structure and worldwide scope of modern terrorist groups makes surveillance essential to uncovering these lethal networks before they strike us in ever more devastating ways.

In one recent terrorism investigation, three defendants were charged with providing material support to terrorists as well as solicitation of terrorist crimes of violence, including kidnapping and murder. Literally all of the evidence against these three defendants consists of audio recordings and fax transmissions obtained through wiretaps and listening devices.

As critical as electronic surveillance is to the investigation of many serious crimes, it is becoming technologically more difficult to carry out wiretap orders and, for some State and local authorities, sometimes impossible to do so. There have been occasions where, because of technological gaps with respect to certain services, telecommunications carriers were unable to provide, or were unable to provide in usable form, the content of communications or related information as required by court orders.

Moreover, criminals and terrorists certainly do not want to be caught. They know that electronic surveillance is an extremely effective law enforcement tool, and they are known to use particular technologies that they suspect law enforcement will have difficulty intercepting.

CALEA's provisions thus are critical to ensuring public safety and national security. CALEA applies to all telecommunications carriers, a term that is specifically defined in the CALEA statute and that is distinct from, and more expansive than, the term "telecommunications carrier" used in the Communications Act of 1934.

[Pause in proceedings.]

Mr. UPTON. Is your sound back now?

The REPORTER. Yes, it is.

Mr. UPTON. Okay. Go ahead.

Ms. PARSKY. CALEA requires telecommunications carriers to be able to execute court-ordered wiretaps by isolating and providing to the government, in real time, the pertinent communications.

Carriers also must have the ability to isolate and provide reasonably available call-identifying information, such as numbers dialed, that is the subject of a pen register or other court order.

CALEA does not allow the government to dictate the design of telecommunications systems, but it does require manufacturers and providers to consult and plan, so that new services that they deploy are CALEA compliant. This approach is appropriate, because any amount of time that a terrorist or other dangerous criminal can use a communications service without a capability for court-ordered interception is too long.

Furthermore, it is important to make clear that CALEA itself actually provides critical protection of privacy rights. The argument that full implementation of CALEA will threaten individual privacy rights is simply misguided. CALEA strikes a delicate balance among three sometimes competing goals:

One, to preserve a narrowly focused capability for law enforcement agencies to carry out properly authorized intercepts; two, to protect privacy in the face of increasingly powerful and personally revealing technologies; and three, to avoid impeding the development of new communications services and technologies.

As the House of Representatives explained in the report, "the bill further protects privacy by requiring the systems of telecommunications carriers to protect communications not authorized to be intercepted."

CALEA addresses privacy concerns in two ways. First, it requires that providers be able to separate out the communications involving the equipment, facilities, or services of the particular subscriber whose communications law enforcement has an order to intercept. This provision promotes both efficiency and privacy.

Second, CALEA requires that a service provider be able to separate out call-identifying information from the content of communications. This protects the call content from law enforcement access where law enforcement only has legal grounds to obtain the call-identifying information. A carrier's compliance with CALEA when implementing a court-ordered wiretap or a pen register order thus protects individuals' privacy rights.

As I have mentioned, the Department of Justice has petitioned the FCC to issue a rulemaking with respect to the application of CALEA to advanced communications technologies, such as broadband Internet access and certain forms of broadband telephony.

It is important to make clear that through this petition to the FCC, the Department is not asking for expansion of CALEA; that is something only Congress is empowered to do.

Rather, we have asked the Commission, pursuant to its mandate, to interpret and implement CALEA in light of emerging telecommunications technologies and an apparent confusion among some service providers and sectors of the telecommunications industry concerning their CALEA obligations.

Mr. UPTON. Ms. Parsky, the clock is not working at the desk, but you have exceeded your time by a considerable amount. I just wonder if you could just summarize, and I will be a little more careful with the clock.

Ms. PARSKY. Certainly. My apologies.

Mr. UPTON. All right.

Ms. PARSKY. Last month the FCC unanimously issued a Notice of Proposed Rulemaking and Declaratory Ruling concerning the issues raised in our petition. Although this is a lengthy document and very complex, we are in the process of fully evaluating it and we will be submitting our formal comments to the FCC soon.

And finally, I would just like to thank you for this opportunity to speak about an issue that is very important to the Department of Justice and important to our efforts to protect national security and public safety.

[The prepared statement of Laura H. Parsky follows:]

PREPARED STATEMENT OF LAURA H. PARSKY, DEPUTY ASSISTANT ATTORNEY GENERAL, CRIMINAL DIVISION, U.S. DEPARTMENT OF JUSTICE

I. INTRODUCTION

Good morning, Chairman Upton, Ranking member Markey, and Members of the Subcommittee. The Department of Justice appreciates the opportunity to address you today on this important subject. As we all are aware, the "Digital Age" in which we now live is offering and will continue to offer tremendous opportunities in telecommunications for both consumers and businesses. The use of high-speed Internet access services is growing rapidly in the United States. In fact, at least one recent report indicates that, for the first time, more U.S. households now connect to the Internet through cable, DSL, and other means of broadband access than through traditional dial-up service. Also, more and more traditional telephone companies, cable companies, and others are offering some means of broadband telephony using Voice over Internet Protocol (VoIP), attracting more and more consumers every day. It is widely believed that such services will essentially replace traditional telephone service in the United States in the not-so-distant future.

The Administration fully supports the rapid and widespread deployment of these communications technologies, understanding that they promise to contribute to increased American productivity and to offer the convenience of reasonably-priced, high-quality service with a variety of useful new features for consumers. Moreover, we welcome and applaud your efforts and the efforts of others in Congress as you carefully debate the proper regulatory environment for new communications technologies. We recognize that we are rapidly expanding into a new and promising world of communications. To automatically apply old-fashioned and likely outdated principles to a new way of doing business is sure to hamper the development of these promising and potent technologies. However, in devising new principles for governing new technologies, we must preserve those safeguards that are critical to our national security and public safety.

The core issue here is responsibility—responsible government and responsible citizenship. By re-evaluating traditional regulation of communications systems, the government is acting responsibly. Likewise, those who develop and provide such communications services must also assume responsibility. The Communications Assistance for Law Enforcement Act (CALEA), 47 U.S.C. 1001, et. seq., was drafted ten years ago when Congress could not have anticipated the details of today's communications revolution. However, Congress did have the foresight to predict that such a communications revolution would take place. CALEA requires that, as new technologies are developed, providers act responsibly by engineering their systems in a way that allows law enforcement to execute court-ordered electronic surveillance.

As communications technology has progressed, some carriers have never questioned their legal obligations under CALEA or their corporate obligations to act responsibly where public safety and national security are at risk. For each and every carrier in this category, we recognize and applaud their leadership and responsibility. Unfortunately, however, there are also some carriers who have deployed their technologies without regard to law enforcement's ability to execute court-ordered electronic surveillance and without regard to their corporate responsibility where public safety and national security are at risk. Because of the existence of carriers in this latter category, we have been forced to petition the Federal Communications Commission (FCC) to affirm the legal obligations of carriers to comply with CALEA and to meet non-compliance with robust enforcement actions.

CALEA's obligations are even more important today than they were when the statute was enacted ten years ago. While many carriers act responsibly and in the public interest without the need for compulsory process, there will always be some businesses that will choose to operate without regard to such concerns. Because savvy criminals and terrorists seek out those businesses, we must take steps to eliminate the vulnerability in our national security and public safety created by those businesses. CALEA and the robust enforcement of CALEA will help accomplish this critical goal.

II. CALEA IS CRITICAL TO ENSURING THAT FEDERAL, STATE, AND LOCAL AUTHORITIES CAN CARRY OUT THE COURT-ORDERED ELECTRONIC SURVEILLANCE THAT IS ESSENTIAL TO THWARTING THE ACTIVITIES OF TERRORISTS AND OTHER SIGNIFICANT CRIMINALS.

CALEA applies to all telecommunications carriers, a term that is specifically defined in the CALEA statute and that is distinct from and more expansive than the term "telecommunications carrier" used in the Communications Act of 1934, 47 U.S.C. 151 et seq. CALEA requires telecommunications carriers to be able to execute court-ordered wiretaps by isolating and providing to the government, in real-time, the pertinent communications. Carriers also must have the ability to isolate and provide reasonably available call-identifying information, such as numbers dialed, that is the subject of a pen register or other court order. New systems and services thus should be developed and deployed, not in a vacuum, but with recognition of law enforcement's legitimate electronic surveillance needs.

CALEA itself does not authorize wiretaps or pen registers. That authority and the requirements for obtaining the relevant court orders are set forth in other statutes. What CALEA does is to help ensure that, as new telecommunications technologies are developed, carriers using those technologies are capable of isolating and providing to the government communications and related information as required by court orders.

When enacting CALEA in 1994, Congress "concluded that there is sufficient evidence justifying legislative action that new and emerging telecommunications technologies pose problems for law enforcement." H.R. Rep. No. 103-827, at. 14. At that time, Congress was especially cognizant of intercept problems associated with the burgeoning wireless industry and the development of custom calling features. Congress, however, anticipated that future technologies would pose similar problems and thus stated that the purpose of the statute "is to preserve the government's ability, pursuant to court order or other lawful authorization, to intercept communications involving advanced technologies . . . or features and services . . . while protecting the privacy of communications and without impeding the introduction of new technologies, features and services." *Id.* at 13.

III. ELECTRONIC SURVEILLANCE IS A CRITICAL LAW ENFORCEMENT TOOL.

It is, of course, no secret that today's criminals use ordinary telephones, cellular telephones, pagers, and the Internet, among other communications devices, in order to coordinate their illicit activities. In investigating terrorism, espionage, and other serious crimes, electronic surveillance is not only one of the most effective tools government has, but often it is the only effective tool. Often criminal organizers and kingpins keep their distance from the criminal conduct they direct through the use of modern communication tools.

There can be no doubt that electronic surveillance takes dangerous criminals off the streets by providing evidence that law enforcement could not have obtained any other way. In fact, one of the requirements for obtaining a federal wiretap order is demonstrating that normal investigative techniques have been or are likely to be inadequate or are too dangerous. Last year alone, 3,674 people were arrested based on evidence obtained through federal and state law enforcement wiretaps. Over the past ten years, over 54,000 people have been arrested based on wiretap evidence. That is as many as 54,000 criminals that might have escaped justice had it not been technologically possible to carry out court-ordered electronic surveillance.

For instance, in a 2002 investigation into members of the Lucchese crime family in New York, wiretaps on cellular telephones and pagers were instrumental in identifying and obtaining convictions of approximately 35 defendants, including three members of the Bonanno crime family. The types of crimes discussed over the wiretapped phones included witness tampering, cocaine distribution, extortion and violence in aid of racketeering, loansharking, and illegal gambling.

In a recent investigation of a marijuana distribution network operating in New York, an intercepted call over a wiretapped phone alerted police to a robbery and double homicide which had just occurred in the Bronx. That valuable evidence allowed authorities to arrest three individuals within hours of the homicides. Investigators later established that several individuals had attempted to rob the targeted marijuana sellers. During the attempted robbery, two individuals were killed by gunshot wounds and a third was shot in the chest and survived. The wiretap evidence helped police piece together the events that had occurred and also helped establish narcotics trafficking charges against additional defendants.

Electronic surveillance is also critical to identifying and ultimately dismantling organized criminal networks, including major national and international drug cartels. Last year, a wiretap in Georgia led to seizures of tons of illegal drugs and millions of dollars. Another wiretap investigation led to *over one hundred arrests* in the United States and abroad and numerous U.S. prosecutions, as law enforcement dismantled an international drug distribution ring responsible for bringing large quantities of heroin and cocaine into the United States from Colombia. Electronic surveillance has allowed us to take cocaine, heroin, methamphetamine, and many other dangerous drugs off our streets and away from our children.

While electronic surveillance remains vital to investigating scourges such as organized crime and drug trafficking, against which we continue to fight, it is even more important to the Department's highest priority—fighting the war on terrorism. The cell structure and worldwide scope of modern terrorist groups make electronic surveillance essential to uncovering these lethal networks before they strike us in ever more devastating ways. In one recent terrorism investigation, three defendants were charged with providing material support to terrorists as well as solicitation of terrorist crimes of violence, including kidnapping and murder. Virtually all of the evidence against these three defendants consists of audio recordings and fax transmissions obtained through wiretaps and listening devices.

Electronic surveillance consistently helps authorities *prevent crimes and save lives.* In a recent child sexual exploitation investigation in Oklahoma, investigators obtained judicial authorization to intercept all wire communications of a pimp who traveled interstate in order to sell children for sexual activity. The pimp was recorded talking about grooming children to become prostitutes, physically beating his victims into compliance, and marketing the children as prostitutes in numerous states. Further, the electronic surveillance helped identify a national child prostitution network and generated investigations in other states. To date, the United States Attorney's Office in Oklahoma City has federally charged nine defendants for sexually exploiting children, and more indictments are pending. Significant state charges have also been filed against ten perpetrators of these horrible crimes. Already, three children (one from Las Vegas, one from New Mexico, and one from Oklahoma) have been rescued by law enforcement thanks to the electronic surveillance. Moreover, probably thousands of physical and sexual assaults upon children have been prevented as a result of these prosecutions that were dependent upon electronic surveillance.

In a narcotics-related wiretap investigation in the New Orleans area, the target of the investigation discussed arrangements for a heroin transaction with traffickers from New York. In subsequent intercepted conversations, the target told his narcotics associate that he intended to kill the New York suppliers after they delivered the heroin. Based upon this information, law enforcement quickly arrested the New York suppliers and thwarted their intended murder. The New Orleans target was then arrested, pleaded guilty, and was ultimately sentenced to life in prison.

In another case, wiretaps used to investigate a violent Russian brigade helped to develop evidence of the organization's involvement in armed robberies, extortion, and arson, among other crimes. Calls intercepted during the investigation uncovered plans for a violent kidnapping-for-ransom scheme. The wiretap evidence allowed law enforcement to quickly make the arrests necessary to prevent the kidnapping.

IV. IN THE ABSENCE OF COMPLIANCE WITH CALEA, TECHNOLOGICAL CONSTRAINTS CAN PREVENT OR HINDER WIRETAPS, ALLOWING CRIMINALS TO EXPLOIT PERCEIVED TECHNOLOGICAL GAPS TO AVOID INTERCEPTION.

As critical as electronic surveillance is to the investigation of many serious crimes, it is becoming technologically more difficult to carry out wiretap orders and, for some state and local authorities, sometimes impossible to do so. There have been occasions where, because of technological gaps with respect to certain services, telecommunications carriers were unable to provide, or were unable to provide in usable form, the content of communications or related information as required by court orders.

Simply put, the equipment needed to carry out an intercept order or pen register has become more sophisticated as telecommunications technology has advanced. Today's digitized communications are provided by many different companies who use many different protocols and transmit communications over many different wires and cables and over a myriad of frequencies through the air— even during a single call. CALEA therefore requires that telecommunications carriers and their equipment vendors work together in designing new technology so that court-ordered interception is technologically possible.

CALEA's provisions are critical to ensuring public safety and national security. Criminals know that electronic surveillance is an extremely effective law enforcement tool, and they have always gone to great lengths to avoid it. Their tactics have included the use of numerous communication devices in order to isolate the damage done if a particular device is compromised and, most relevant to CALEA, the quick migration to particular technologies that they suspect law enforcement will have difficulty intercepting. Criminals and terrorists certainly do not want to be caught, and they are quick to take advantage of any perceived gap in our ability to detect and disrupt their criminal activities.

V. THE FCC IS CAREFULLY CONSIDERING THE APPLICATION OF CALEA TO ADVANCED TELECOMMUNICATIONS TECHNOLOGIES.

In the face of the real and growing threat to public safety and national security posed by the misuse of VoIP and other new telecommunications technologies, the Department of Justice has petitioned the FCC to issue a rulemaking with respect to the application of CALEA to advanced communications technologies such as broadband Internet access and certain forms of broadband telephony. This subcommittee hearing comes in the midst of the FCC's consideration of the Department's petition and the resulting, vibrant discourse involving the Department, other law enforcement entities, industry, and special interest groups.

In our petition for expedited rulemaking, filed last March, we requested that the Commission rule that CALEA applies to broadband internet access services and certain forms of broadband telephony services; reaffirm that the push-to-talk services now offered by many cellular telephone companies are subject to CALEA; identify the packet-mode services covered by a CALEA implementation Order issued in 1999 and establish compliance deadlines with respect to that Order; adopt rules for expeditiously determining whether a new technology is subject to CALEA and for establishing compliance deadlines and administrative enforcement procedures for noncompliance; and resolve cost recovery issues.

It is important to make clear that through this petition to the FCC, the Department is not asking for expansion of CALEA; that is something only Congress is empowered to do. Rather, we have asked the Commission, pursuant to its mandate, to interpret and implement CALEA in light of emerging telecommunications technologies and an apparent confusion among some service providers and sectors of the telecommunications industry concerning their CALEA obligations.

In crafting CALEA, Congress wisely did not limit its scope to one particular technology, service, or suite of features, but rather set in place a structure that anticipated and provided for a vast array of technological advances. As the then Director of the FBI testified in support of the legislation, CALEA was

> intended to stand the test of time.... It is specifically designed to deal intelligently and comprehensively with current and emerging telecommunications technologies and to preclude the need for much more restrictive and more costly legislation in five or ten years when court-authorized interceptions would no longer be possible due to further technology advances.

Hearing on Police Access to Advanced Communications Systems Before the Senate Subcommittee on Technology and the Law of the Committee on the Judiciary and the House Subcommittee on Civil and Constitutional Rights of the Committee on the Judiciary (statement of Louis J. Freeh, Director of the Federal Bureau of Investigation). Thus, Congress has already recognized the importance of ensuring that, as advanced communications technologies develop, industry develops the technical means to implement court orders.

In response to the Department's petition, dozens of state and local law enforcement entities and associations filed comments with the FCC emphasizing the critical need to preserve CALEA. State and local entities conduct annually almost three-fifths of all wiretaps in the United States. As articulately expressed by the National Association of District Attorneys:

> For over a decade we have been pleading for the tools and the laws we need to protect the people in our communities. We will never know whether we could have prevented the tragic consequences of September 11th had we had the investigative tools we have been asking for since 1992. We only know that we will need every advantage to prevent such a tragedy from ever occurring again.

Comments of the National Association of District Attorneys, In the Matter of Joint Petition for Rulemaking to Resolve Various Outstanding Issues Concerning the Implementation of the Communications Assistance for Law Enforcement Act, FCC 04-187, at 2.

Moreover, many of the responsible members of the communications industry have agreed with law enforcement, through comments filed in other related proceedings, that carriers play an important role in protecting public safety and national security. One industry association put it simply: "American citizens should be assured that communications companies are providing appropriate help to law enforcement." Comments of the United States Telecommunications Association, In the Matter of IP-Enabled Services: Notice of Proposed Rulemaking, FCC 04-28, at 36-37.

VI. IN ITS RECENT NPRM, THE FCC HAS RECOGNIZED THE IMPORTANCE OF CALEA IN THE CONTEXT OF EMERGING ADVANCED TECHNOLOGIES.

Last month, after receiving extensive comments on the Department's petition, the FCC unanimously issued a Notice of Proposed Rulemaking and Declaratory Ruling concerning a wide variety of CALEA issues ("CALEA NPRM"). The CALEA NPRM states unequivocally that "it is the Commission's primary policy goal to ensure that [law enforcement agencies] have all of the resources that CALEA authorizes to combat crime and support Homeland Security," and it recognizes the need to balance that interest with the competing privacy and technology development interests. CALEA NPRM at ¶4. While the Department is still analyzing this lengthy issuance and will soon provide formal comments to the FCC, a few things are important to highlight. The CALEA NPRM tentatively concludes that CALEA applies to such services as facilities-based broadband Internet services and managed VoIP telephone services, seeking comment on the FCC's legal reasoning to support such conclusions. In addition, the Commission issued a declaratory ruling that wireless push-to-talk services are subject to CALEA. Although the Commission did not agree with the Department on every point raised in our petition, we are pleased with the seriousness with which the Commission is approaching these critical issues.

Further, in the CALEA NPRM, the FCC recognized that law enforcement does not seek the power to dictate how the Internet should be engineered or the power to veto the deployment of new telecommunications services. Law enforcement cannot—nor do we seek to—dictate to any carrier how best to design its service or what services it can or cannot offer. We only ask that any service comply with the law in order not to imperil public safety and national security. In light of the fact that CALEA solutions can be just as innovative as the services themselves, the FCC appropriately committed itself to "finding solutions that will allow carriers and manufacturers to find innovative ways to meet the needs of the law enforcement community without adversely affecting the dynamic telecommunications industry." CALEA NPRM at ¶31.

It is worth noting that nothing in the CALEA NPRM precludes the FCC from making an independent assessment of whether a carrier is subject to other economic regulation under the Communications Act of 1934, as amended. In confining its analysis to CALEA, the Commission explicitly stressed that the CALEA NPRM "in no way predispose[s] how the Commission may proceed with respect to adopting a regulatory framework for Internet Protocol ("IP")-enabled or broadband services or determining their legal classification under the Communications Act." CALEA NPRM at ¶1, n. 1.

VII. SEVERAL MISCONCEPTIONS ABOUT CALEA AND THE DEPARTMENT'S EFFORTS TO SECURE ITS IMPLEMENTATION WARRANT CLARIFICATION.

I'd like to take a few moments to address several misconceptions about CALEA and about the Department's implementation efforts.

A. The Department's Petition Does Not Seek to Erode the Strict Constitutional, Statutory and Regulatory Limitations Imposed on Electronic Surveillance.

While electronic surveillance is a necessary tool, we are mindful that it is also a very powerful tool—one that has serious implications for the privacy of citizens. Accordingly, law enforcement only uses electronic surveillance as a method of last resort, and even then we adhere to strict limitations on its use.

As I briefly mentioned before, CALEA itself does not authorize electronic surveillance. In presenting our views to the FCC concerning the interpretation of CALEA, the Department is not seeking expanded authority to conduct wiretaps. As Congress said when enacting CALEA, "[s]ince 1968, the law of this nation has authorized law enforcement agencies to conduct wiretaps pursuant to court order. That authority extends to voice, data, fax, e-mail and any other form of electronic communication. The bill will not expand that authority." H.R. Rep. No. 103-827, at 17.

The limitations on law enforcement's use of wiretaps are imposed by the Constitution, statutes, and internal Department procedures. First, the U.S. Constitution obviously places important parameters on our use of electronic surveillance. Under the Fourth Amendment, the government must demonstrate probable cause to a neutral magistrate before obtaining a warrant for a search, arrest, or other significant intrusion on privacy.

Congress and the courts have also provided statutory limits beyond those required by the Constitution. For instance, law enforcement must obtain a "trap and trace" or "pen register" court order to obtain information identifying who is receiving or sending communications to or from a particular suspect, even though not required under the Constitution. *See* 18 U.S.C. 3121 et. seq.

The statutory authorization for law enforcement wiretaps, 18 U.S.C. §§2510-22 (commonly known as "Title III"), as amended by the Electronic Communications Privacy Act (ECPA) in 1986, creates an even higher burden for obtaining the real-time interception of the content of communications. The Senate Report on Title III stated explicitly that the legislation "has as its dual purpose (1) protecting the privacy of wire and oral communications and (2) delineating on a uniform basis the circumstances and conditions under which the interception of wire and oral communications may be authorized." Senate Committee on the Judiciary, Omnibus Crime Control and Safe Streets Act of 1967, S. Rep. No. 1097, 90th Cong., 2d Sess. (1968) at 66. When Title III was updated in 1986 to include provisions regarding electronic communications, the Senate Report stated that ECPA represented "a fair balance between the privacy expectations of American citizens and the legitimate needs of law enforcement agencies." Senate Committee on the Judiciary, Electronic Communications Privacy Act of 1986, S. Rep. No. 541, 99th Cong., 2d Sess. (1986) at 5. Accordingly, under Title III, in order to obtain a court order to capture the contents of communications as they occur, the government must show that normal investigative techniques for obtaining information about a serious felony offense have been or are likely to be inadequate or are too dangerous, and that any interception will be conducted so as to ensure that the intrusion is minimized.

Even beyond the limits placed by the Constitution and the Congress, the Department of Justice has its own internal procedures to provide still more safeguards. For example, the Office of Enforcement Operations (OEO) in the Criminal Division of the Department reviews proposed Title III applications to ensure that the request for interception satisfies the protections of the Fourth Amendment and complies with applicable statutes and regulations. Even if OEO recommends authorizing a request, the application cannot go to a court without approval by a Deputy Assistant Attorney General or higher-level official in the Department. The fact that not a single application for electronic surveillance under Title III was rejected by a federal court in all of 2003 is a testament to the vigilance and care the Department takes when asking for this authority.

If the Department of Justice approves a federal Title III request, it still must, of course, be submitted to and approved by a court of proper jurisdiction. The court will evaluate the application under the Fourth Amendment and using the familiar standards of Title III. By statute, for example, the application to the court must show, through sworn affidavit, why the intercept is necessary as opposed to other less-intrusive investigative techniques. The application must also provide additional detail, including whether there have been previous interceptions of communications of the target, the identity of the target (if known), the nature and location of the communications facilities, and a description of the type of communications sought and the offenses to which the communications relate. By statute and internal Department regulation, the interception may last no longer than 30 days without an extension by the court. All intercepted communications are sealed by the court, further protecting privacy.

Often courts also impose their own safeguards. For example, many federal courts require that the investigators provide periodic reports to the court setting forth information such as the number of communications intercepted, the steps taken to minimize irrelevant traffic, and whether the interceptions have provided information relevant to the criminal investigation. The court may, of course, terminate the interception at any time.

It is only after we have complied with these comprehensive regulatory, statutory, and Constitutional protections that CALEA comes into play and ensures that a court order can be implemented. Our recent filings with the FCC do not seek to change any part of this carefully calibrated system.

B. Implementation of CALEA Will Help Protect Privacy.

It is important to make clear that CALEA, itself, actually provides critical protection of privacy rights. The argument that full implementation of CALEA will threaten individual privacy rights is simply misguided. CALEA strikes a delicate balance among three sometimes competing goals: "(1) to preserve a narrowly focused capability for law enforcement agencies to carry out properly authorized intercepts; (2) to protect privacy in the face of increasingly powerful and personally revealing technologies; and (3) to avoid impeding the development of new communications services and technologies." H.R. Rep. No. 103-827, at 13. As the House of Representatives explained in the report, "the bill further protects privacy by requiring the systems of telecommunications carriers to protect communications not authorized to be intercepted." *Id.* at 10.

CALEA addresses privacy concerns in two ways. First, it requires that providers be able to separate out the communications involving the equipment, facilities, or services of the particular subscriber whose communications law enforcement has an order to intercept. This provision promotes both efficiency and privacy. Second, CALEA requires that a service provider be able to separate out call-identifying information from the content of communications. This protects the call content from law enforcement access where law enforcement only has legal grounds to obtain the call-identifying information. CALEA Section 103; 47 U.S.C. 1002. A carrier's compliance with CALEA when implementing a court-ordered wiretap or a pen register order thus protects individuals' privacy rights.

C. In Keeping with the Provisions of CALEA, the Department of Justice Does not Seek to Dictate the Design of Telecommunications Systems.

It is also important to stress that the Department does not seek to dictate the design of new telecommunications systems. In fact, CALEA explicitly prohibits any such undertaking by providing that it "does not authorize any law enforcement agency or officer . . . to require any specific design . . . to be adopted by any provider [or] manufacturer . . . ," and it does not authorize any law enforcement agency or officer "to prohibit the adoption of any equipment, facility, service, or feature by any provider . . . [or] manufacturer." CALEA Section 103, 47 U.S.C. 1002(b)(1).

What the Department does seek is to ensure that new communications services and features to which CALEA applies are deployed with CALEA solutions in place whenever feasible. Indeed, Section 106 of CALEA mandates that carriers consult with manufacturers "as necessary, in a timely fashion" to ensure "that current and *planned* equipment, facilities, and services comply with [CALEA] capability requirements[.]" 47 U.S.C. 1005 (emphasis added). CALEA solutions may be developed by individual service providers or by industry, but they must be developed. Any amount of time that a terrorist or other dangerous criminal can use a communications service without a capability for court-ordered interception is too long.

D. The Department is Not Seeking to Re-allocate the Costs of CALEA Implementation.

Finally, the Department is not seeking to re-allocate the costs of CALEA implementation to industry or consumers. It is CALEA itself that places any cost burden on telecommunications carriers in the first instance, rather than on the government, for equipment, facilities, and services installed or deployed after January 1, 1995. CALEA Section 109(b); 47 U.S.C. 1008(b). This same provision, however, also allows carriers to seek a determination of whether implementation of a CALEA solution is "reasonably achievable" in light of costs and other issues and allows carriers to seek compensation for costs or reprieve in some circumstances. CALEA recognizes that the greatest cost efficiency can usually be achieved by building intercept solutions into a system's initial design prior to deployment, rather than as a retrofit.

VIII. CONCLUSION

Now, ten years after the enactment of CALEA, we must not back away from the important principles behind CALEA. If anything, it is even more critical today than in 1994 that advances in communications technology not provide a haven for criminal and terrorist activity. While we recognize the desirability of and need for the development and deployment of advanced telecommunications technologies, we must at the same time act responsibly to preserve the national security and public safety mandates of CALEA. The Department of Justice appreciates this Subcommittee's leadership in seeking to promote new telecommunications technologies in a manner that addresses these national security interests, and we thank you for your continuing support.

Mr. UPTON. Thank you. Thank you very much.
Mr. Thomas.

STATEMENT OF MARCUS C. THOMAS

Mr. THOMAS. Thank you.

Good morning, Chairman Upton, members of the subcommittee. I am grateful for this opportunity to discuss this important national security and public safety issue, law enforcement's access to communication systems in the digital age.

Let me say up front that I believe it's important to state that the FBI and the law enforcement community recognize the importance of the continued development and adoption of innovative technologies to insure that the United States remains a leader in today's competitive global marketplace. I believe that public safety, national security, and technology innovation can all be served by good policy.

I also do not think anyone seriously challenges the need for law enforcement and national security communities to be able to conduct court authorized electronic surveillance. There is no doubt that wiretaps produce powerful intelligence and evidence against the most dangerous criminals and terrorists. When police cannot use other investigative techniques to safely and successfully collect evidence and intelligence, they often use wiretaps to catch criminals with words uttered from their own mouths.

Concerns regarding the serious threat to our capabilities are not limited to the United States law enforcement and national security communities. Worldwide new laws are being implemented that are intended to require network providers to furnish communications interception services to government agencies.

The technical assistance of communications service providers in helping law enforcement agencies to execute an electronic surveillance order is always important, and in many cases it's absolutely essential. This circumstance has proven to be the case increasingly with the advent over the past 10 years or so of complex new systems, services and features.

In the House report accompanying CALEA when it was passed in 1994, the purpose of the legislation was clearly set out to make clear telecommunications carriers' duty to cooperate in the interception of communications for law enforcement purposes.

In short, CALEA's intent was to mandate through service provider cooperation access where advancing technology would otherwise preclude it.

Despite the fact that since the enactment of CALEA there have been technological advancements that few of us could have foreseen, the implementation of CALEA has been successful. Referring to the most recent wiretap report published annually by the Administrative Office of the United States Courts, more than 70 percent of all criminal wiretap authorizations listed were through CALEA compliant capabilities.

In recent years, the FBI has found that there are greater and more diverse challenges in effectuating the electronic surveillance orders within modern networks than with conventional telephony networks operated by traditional telecommunications carriers.

In order to implement electronic surveillance orders in these diverse networks, the FBI has relied on elaborate and costly technical approaches to insure that only messages for which there's probable cause to intercept are, in fact, intercepted, and that all such authorized messages are intercepted.

As a result, it has become increasingly common for the FBI to seek and for judges to issue orders for Title III or FISA interceptions which are much more complex and detailed and much more likely to be directed to multiple network operators and service providers than earlier orders which are ordinarily directed against a single plain, old telephone service provider.

The issue that I have described may be too complex for one remedy to solve. Like so many issues we try to deal with today, the future success of law enforcement electronic surveillance will depend upon a multi-pronged approach. In response to the challenges presented by rapid technological advances, the FBI and law enforcement community have been using all available means to implement their mission, to protect national security and public safety.

In my written testimony, I included a list of significant issues which we are addressing, including technology advancements, industry cooperation, third party services, industry standards and specifications, law enforcement coordination and costs. I would encourage the subcommittee and the rest of the members discussing these issues to keep in mind the need for continued access by U.S. law enforcement to our Nation's communications infrastructures.

Experience has proven that statutorily imposed responsibilities must necessarily be one element of the solution, but not the only element. As such, we must continue to have statutory mandates such as CALEA and build upon them using varied tools, including incentives.

In conclusion, I'd like to say over the past 10 years or more, we have witnessed continued steady growth in computer and Internet related crimes, including extremely serious acts in furtherance of terrorism, espionage, infrastructure attack, as well as more conventional serious and violent crimes.

These activities, which even now are being planned and carried out, in part using the Internet and other complex networks and services, pose challenges to the national security and law enforcement communities that we dare not fail to meet.

In turn, the ability of the FBI and law enforcement community to effectively investigate and prevent these serious crimes is, in part, dependent upon our ability to lawfully and effectively intercept and acquire vital intelligence and evidence of crimes and our ability to promptly respond to these threats to the American public. As the networks become more complex so does the challenge placed upon us to keep pace.

I look forward to working with the subcommittee staff to provide more information and welcome your suggestions to this important national security and public safety issue. Thank you for including my written testimony in the record, and I'll be happy to answer questions.

Thank you.

[The prepared statement of Marcus C. Thomas follows:]

PREPARED STATEMENT OF MARCUS C. THOMAS, DEPUTY ASSISTANT DIRECTOR, INVESTIGATIVE TECHNOLOGY DIVISION, FEDERAL BUREAU OF INVESTIGATION

Good morning, Chairman Upton, Ranking Member Markey, and Members of the Subcommittee, I am grateful for this opportunity to discuss this important national security and public safety issue: law enforcement's access to communications systems in the digital age. I would like to start by briefly outlining a historical framework of court-authorized electronic surveillance in highly-complex communications networks, then discussing the situation in which the law enforcement community currently finds itself, and some of the problems with which we are currently dealing. Lastly, I would like to briefly discuss some of our ongoing efforts intended to address a number of these problems.

BACKGROUND

Prior to delving into the subject of electronic surveillance, I believe it is important to state that the FBI and the law enforcement community recognize the importance of the continued development and consumer adoption of innovative technologies to ensure the United States remains a leader in today's competitive, global marketplace. One of the fundamental requirements for preserving national security, the privacy of our citizens, and public safety is ensuring that United States national security and law enforcement agencies are able to securely and effectively use lawful process to gather evidence and intelligence during investigations. We remain extremely concerned about the very serious, public safety and national security threat posed by the misuse of technologies that hamper lawfully-authorized electronic surveillance of communications occurring over their systems. I believe that public safety, national security, and technological innovations can be served by good policy. .

I do not think anyone seriously challenges the need for the law enforcement and national security communities to be able to conduct court-authorized electronic surveillance. There is no doubt wiretaps produce powerful intelligence and evidence against the most dangerous criminals and terrorists. When police cannot use other investigative techniques to safely and successfully collect evidence and intelligence, they often use wiretaps to catch and convict criminals with words uttered from their own mouths. Concerns regarding this serious threat are not limited to the United States law enforcement and national security communities. Worldwide, new laws are being implemented that are intended to require network providers to furnish communications interception services to government agencies.

The issue I have just described may be too complex for one remedy to solve. Like so many issues we try to deal with today, the future success of lawful electronic surveillance will depend on a multi-pronged approach. In some instances, responsibilities mandated of a service provider are the appropriate course of action. In others, to meet the exigent needs of law enforcement, industry cooperation can be the most constructive avenue of pursuit. Finally, any approach would be incomplete without considering law enforcement's own abilities. I am here today, mere days before the third anniversary of September 11th, to stress the importance of the outcome of our discussion: *law enforcement's continued ability to conduct lawful electronic surveillance to ensure national security and public safety.*

TECHNICAL ASSISTANCE REQUIREMENTS

As the Subcommittee is aware, there are two federal statutory regimens pertaining to electronic surveillance one regarding criminal investigations; the other regarding foreign intelligence, counterintelligence, and terrorism investigations. The former is derived from Title III of the Omnibus Crime Control and Safe Streets Act of 1968 (commonly referred to as "Title III"), as amended, and portions of the Electronic Communications Privacy Act of 1986 (ECPA), as amended. The latter is derived from the Foreign Intelligence Surveillance Act of 1978 (FISA), as amended. Regardless of the statutory regimen, Congress took action in 1994 to mandate telecommunications carriers, and others as identified by the FCC, to ensure their networks were capable of conducting electronic surveillance.

The technical assistance of communications service providers in helping a law enforcement agency execute an electronic surveillance order is always important, and in many cases it is absolutely essential. This circumstance has proven to be the case increasingly with the advent, over the past ten years or so, of advanced communications services and features. Accordingly, Title III and FISA, as well as most state electronic surveillance laws, mandate service provider assistance incidental to law enforcement's execution of electronic surveillance orders.

Title III specifies that a "service provider, landlord, custodian, or other person shall furnish the applicant forthwith all information, facilities, and technical assistance necessary to accomplish the interception unobtrusively and with a minimum of interference . . ." upon the request of the applicant (specifically, law enforcement). In practice, judges sign two orders: one order authorizing the law enforcement agency to conduct the electronic surveillance, and a second (abbreviated) assistance order directed to the service provider specifying, for example, the telephone number(s) of the subject that are the object of the order and directing the provision of necessary assistance.

Historically, assistance sought by law enforcement agencies was rather straightforward and basic. For example, law enforcement agencies sought and received service provider assistance to identify line appearance information (i.e., locating the physical appearance of a subject's line) and to establish leased lines running from the point of interception to a monitoring facility of the law enforcement agency. This model was very effective prior to the advent of advanced calling features and the introduction of mobile communications. Likewise, law enforcement agencies have historically paid reasonable expenses for such administrative assistance.

In 1994, as a result of the emergence of an ever increasing array of new services and features, many of which would have impeded, if not precluded, normal electronic surveillance efforts by obstructing lawful access, Congress passed, and the President signed into law, the aforementioned CALEA legislation. In the House Report accompanying CALEA, the purpose of the legislation was clearly identified: "to make clear a telecommunications carrier's duty to cooperate in the interception of communications for law enforcement purposes...". That is to say that a primary purpose of CALEA was to clarify and strengthen the statutory requirement that service providers furnish *"all"* technical assistance necessary to accomplish the interception—meaning to design and build into their networks the capability and capacity requirements needed by law enforcement. It is not enough just to be willing to assist; rather, service providers must actually be capable of making that assistance possible in a rapidly changing technological world. In short, CALEA's intent was to mandate access where advancing technology would otherwise preclude it.

Despite the fact that in the years since the enactment of CALEA there have been technological advancements few of us could have foreseen, CALEA has proven essential to law enforcement successes. In the most recent Wiretap Report (published annually by the Administrative Office of the United States Courts), 80 percent of wiretap authorizations were for cellular or mobile telephones. Of that number, I am pleased to tell you approximately 90 percent were conducted using technical solutions developed specifically in response to the assistance capability requirements identified in CALEA. In other words, more than 70 percent of all criminal wiretap authorizations were "CALEA-compliant." Looking to the future, our success with CALEA's application to cellular telephones can be seen as a model. Prior to the passage of CALEA the 1991 Wiretap Report identified that cellular phones accounted for approximately one percent of wiretap authorizations. CALEA provided a framework to ensure law enforcement's lawful access as criminals migrated to the new technology. I believe we are at the point with Voice over Internet Protocol (VoIP) today that we were with cellular telephones in the early 1990s—with one significant difference: all service providers, both wireline and wireless, have an incentive to migrate their networks to an IP platform. What that means is the transition to a VoIP infrastructure is occurring very quickly. In recognition of this rapid change, we have petitioned the Federal Communications Commission to make clear that CALEA applies to certain forms of I.P. telephony services. We feel this is critical to protecting law enforcement interests.

It is important to note that the requirement for service provider assistance under 18 U.S.C. 2518(4) remains in full force and effect, notwithstanding the applicability of CALEA, and requires service providers to do whatever reasonably can be done to comply with assistance court orders issued by judges. In other words, even when CALEA does not apply, the service provider (or "landlord, custodian, or other person") served with a court order for surveillance is legally required to do whatever can reasonably be done to implement the order.

CURRENT TECHNOLOGY AND POLICY ISSUES

Perhaps the most significant technological challenges in the area of electronic surveillance faced by the law enforcement and national security communities have been those challenges brought on by convergence. Convergence refers to the blurring of lines among traditionally distinct communications products, services, and regulatory structures and can be thought of as the ability (technically and legally) of different network platforms to carry essentially the same kinds of services (so-called network-independence) as well as the ability of a single network platform to carry many different kinds of services (so-called service-independence). Such network/service independence is perhaps most evident in the blurring of wireless and wireline network services, but also in the blurring of data and voice services. The most relevant instrument of change with regard to such convergence has been the emergence of IP networks.

In recent years, the FBI has found that there are greater and more diverse challenges in effectuating electronic surveillance orders within modern networks than with "conventional" telephony networks operated by traditional telecommunications carriers. In order to implement electronic surveillance orders in these diverse networks, the FBI has relied on elaborate and costly technical approaches to ensure that only messages for which there is probable cause to intercept are, in fact, intercepted and that all such authorized messages are intercepted. As a result, it has become increasingly common for the FBI to seek, and for judges to issue, orders for Title III or FISA interceptions which are much more complex and detailed, and much more likely to be directed to multiple network operators and service providers, than earlier orders, which were ordinarily directed against a single "plain old telephone services" provider.

It is important to point out that, when CALEA was passed in 1994, the Internet was a nascent consumer technology, the World Wide Web was only really coming into existence in the laboratory, and wireless telephones were largely voice-only devices and not the web-enabled devices we see today. Nevertheless, the Congress, with CALEA, was attempting to address the complex and varied communications services that we now see.

LAW ENFORCEMENT RESPONSE

In response to the challenges presented by rapid technological advances, law enforcement has been using all available means to implement its mission to protect national security and public safety. First, law enforcement has sought to ensure compliance with CALEA. In keeping with the spirit of Congress's intent when enacting CALEA, the FBI has not sought to apply its requirements either recklessly or broadly to those to whom it should not apply. Because neither CALEA, nor any other single approach, is viewed as the absolute solution for law enforcement's electronic surveillance problems, the FBI and other law enforcement agencies have worked continually to augment CALEA requirements with government capabilities. In this regard, we have worked to develop close liaison relationships with the Information Technology industry as a means of addressing the public safety and national security issues associated with electronic surveillance and the use of technologies which tend to hamper our legitimate interception efforts. Over the past several years, we have been aggressively pursuing an industry outreach strategy to inform the Information Technology industry of law enforcement's needs in the area of electronic surveillance, to continue to encourage the development of interception capabilities that meet law enforcement's needs, and to seek industry's assistance regarding the development of law enforcement tools and capabilities when complex technologies are encountered during the course of lawful investigation. As a result of this strategy, we have seen a number of significant advancements which should be further pursued and emulated.

First, we have seen a number of technological developments which have led to the marketing of comprehensive technical tools designed, in part, to perform electronic surveillance within the complex environment of the Internet. These tools, which are designed to be implemented and operated by a service provider, have greatly extended the capability to effectuate lawful electronic surveillance on ISP networks. Several companies have aggressively developed and marketed such tools.

Second, the FBI and the law enforcement community have always, as a first instinct, sought to work cooperatively and closely with computer network service providers and their software and equipment manufacturers to develop lawful interception capabilities, especially where legal, evidentiary, and investigative imperatives require special purpose tools. As a result, a number of network operators and service providers have acquired and implemented lawful interception capabilities.

Third, we have seen the emergence of so-called "third-party services"—companies, largely utilizing the tools mentioned above, marketing electronic surveillance services to both the network operator community and the law enforcement community. One such third party service provider provides telecommunications network operators, cable operators, and ISPs with a streamlined service to help meet requirements for assisting government agencies with lawful interception and subpoena requests for subscriber records. With respect to third-party service providers, law enforcement sees them as one potential avenue for telecommunications network operators, cable operators, and ISPs to meet their obligations under Title III and/or FISA. Employing a third party may, for example, make a service provider's processes more efficient, but in no way should be seen as relieving the service provider of its electronic surveillance obligations. I liken third-party services to other out-sourced services such as payroll administration, where the third party handles the paperwork, but the buck stops with the company that pays the bill.

Fourth, we have seen a truly commendable effort on the part of CableLabs, an industry trade consortium representing many cable companies, along with Time-Warner, Comcast, CableVision and Cox Communications, to develop and publish a set of technical standards which, on their face, meet law enforcement needs with regard to electronic surveillance capabilities. This standard was developed in a spirit of cooperation which began by recognizing the legitimacy of law enforcement's needs and duties and the unique position industry is in to ensure that our public safety and national security missions are fulfilled.

Fifth, as always, we have seen the law enforcement community pull together in the face of this issue. Speaking for the FBI, I can say that many of our technologies, systems, and processes developed for our own use have been made available, to the extent possible, to the greater law enforcement community, including other federal law enforcement agencies as well as state and local agencies. Nonetheless, the challenges are daunting, and the federal government cannot shoulder this burden alone. Even with federal assistance, state and local law enforcement are currently having significant problems effectuating their interception orders, and the situation will only grow worse.

Finally, another important issue regarding lawful interception which must be addressed is that of cost. One inescapable fact is that lawful electronic surveillance in this modern "digital age" is increasingly complex and rapidly changing. Both of these circumstances have the effect of increasing the overall cost of electronic surveillance. Unfortunately, on this issue, there is no returning to the "days of old" where policemen hunkered down in panel vans on the street corner recording wiretaps on reel-to-reel tape. For now, and for evermore, there is a new baseline for costs associated with this work.

I will leave you with a last thought regarding the capability of law enforcement agencies to lawfully access communications in a "digital age," and that is this: without the "high tech" industry assisting the government in this effort, our challenge will be greater. Law enforcement must have the continued ability to cost-effectively conduct lawful electronic surveillance to ensure national security and public safety. As I mentioned earlier, this is a complex issue that needs a multi-pronged solution. Industry must be engaged and must involve itself in that solution. I would encourage this Subcommittee and the rest of Congress, when discussing the issue, to keep in mind the need for continued access by U.S. law enforcement to our nation's communications infrastructures. Experience has proven that statutorily-imposed responsibilities must necessarily be one element of the solution but not the only element. As such, we must continue to have statutory mandates such as CALEA and build on them, using varied tools, including incentives.

In conclusion, I would like to say that over the last ten years or more, we have witnessed continuing, steady growth in computer and Internet-related crimes, including extremely serious acts in furtherance of terrorism, espionage, infrastructure attack, as well as more conventional serious and violent crimes. These activities which even now are being planned or carried out, in part using the Internet and other complex networks and services, pose challenges to the national security and law enforcement communities that we dare not fail to meet. In turn, the ability of the FBI and the law enforcement community to effectively investigate and prevent these serious crimes is, in part, dependent upon our ability to lawfully and effectively intercept and acquire vital intelligence and evidence of crimes and our ability to promptly respond to these threats to the American public. As the networks become more complex, so too does the challenge placed upon us to keep pace.

I look forward to working with the Subcommittee staff to provide more information and welcome your suggestions on this important national security and public safety issue: law enforcement's access to communications systems in the digital age. I will be happy to answer any questions that you may have. Thank you.

Mr. UPTON. Thank you.

Mr. Knapp.

STATEMENT OF JULIUS P. KNAPP

Mr. KNAPP. Chairman Upton, members of the subcommittee, good morning. I welcome this opportunity to discuss the FCC's activities to implement the Communications Assistance for Law Enforcement Act, or CALEA, for short.

The FCC is strongly committed to insuring the telecommunications carriers provide law enforcement agencies with the surveillance capabilities that are required under CALEA. We recognize the vital importance of lawfully authorized surveillance in combatting crime and insuring homeland security.

The FCC also recognizes that in providing these capabilities, we must not compromise other important objectives, such as avoiding impediments to new technologies and services, protecting personal privacy, and minimizing the impact on consumers.

The CALEA statute was passed in 1994 with the purpose of preserving the government's ability pursuant to court order or other lawful authorization to intercept communications involving advanced technologies, such as digital or wireless transmission modes. Great changes in technology have occurred over the past 10 years which have challenged the ability of law enforcement to conduct lawfully authorized surveillance.

Most notably, there has been a rapid shift from circuit mode to packet mode technologies with an array of new services, such as broadband Internet access and Voice Over Internet Protocol, or VoIP, now offered to businesses and consumers.

The FCC has been proud to facilitate this communications revolution by minimally regulating these new services to promote increased competition in the introduction of new services for businesses and consumers. These changes from a circuit based to a packet based world will have a profound effect on the way we communicate.

However, in the midst of this communications revolution, there has been an upsurge in dangerous criminal activity, including terrorism. Accordingly, the FCC must insure that CALEA's intent is carried out and that lawfully authorized electronic surveillance is not compromised by new technologies.

On August 4, 2004, the FCC adopted a notice of proposed rulemaking and declaratory ruling to launch a thorough examination of the appropriate legal and policy framework for implementing CALEA. This proceeding was initiated in response to a joint petition filed by the Department of Justice, Federal Bureau of Investigation, and Drug Enforcement Administration in March 2004.

These parties state that several issues require immediate attention and resolution by the FCC so that industry and law enforcement have clear guidance as CALEA implementation moves forward.

The notice of proposed rulemaking, or "notice" for short, addresses a number of areas, including the applicability of CALEA to broadband Internet access and VoIP, capability requirements and solutions, compliance extensions, and cost and cost recovery issues.

The notice tentatively concludes that CALEA's provisions apply to facilitates based providers of any type of broadband Internet access service, including wire line, cable modem, satellite wireless, and power line, and to managed or mediated voice over Internet protocol service. The notice finds that these services fall under CALEA as a replacement for a substantial portion of the local telephone exchange service.

The notice also solicits comment on what would be a reasonable amount of time for entities that heretofore have not been subject to CALEA to comply with its requirements.

The companion declaratory ruling clarifies that commercial wireless push-to-talk services are subject to CALEA regardless of the technologies that providers choose to use in offering them.

As Chairman Powell noted in his statement on the CALEA notice, our support for law enforcement is unwavering. The FCC looks forward to developing a complete and comprehensive record before determining how to best proceed. We will devote the necessary resources to expeditiously and responsibly complete this task.

I would like to thank you, Mr. Chairman, for the opportunity to appear before you today. This concludes my testimony, and I would be pleased to answer any questions you or this committee may have.

Thank you.

[The prepared statement of Julius P. Knapp follows:]

PREPARED STATEMENT OF JULIUS P. KNAPP, DEPUTY CHIEF, OFFICE OF ENGINEERING AND TECHNOLOGY, FEDERAL COMMUNICATIONS COMMISSION

Mr. Chairman, Ranking Member, and Members of the Subcommittee: Good morning. I am Julius Knapp, Deputy Chief of the Office of Engineering and Technology at the Federal Communications Commission (FCC or Commission). I welcome this opportunity to discuss the FCC's activities to implement the Communications Assistance for Law Enforcement Act (CALEA).

The FCC, under Chairman Powell's leadership, is absolutely committed to ensuring that telecommunications carriers provide law enforcement agencies (LEAs) with the surveillance capabilities that are required under CALEA. The Commission recognizes the vital importance of lawfully authorized surveillance in combating crime and ensuring Homeland Security and intends for our recently initiated proceeding to continue this ability. The FCC also recognizes that in providing these capabilities we cannot compromise other important objectives, such as avoiding impediments to new technologies and services, protecting personal privacy, and minimizing the impact on consumers.

INTRODUCTION

Since 1970, telecommunications carriers have been required to cooperate with LEAs to assist their conduct of electronic surveillance. The CALEA statute was passed in 1994 with the purpose of preserving the government's ability, pursuant to court order or other lawful authorization, to intercept communications involving advanced technologies such as digital or wireless transmission modes, while protecting the privacy of communications and without impeding the introduction of new technologies, features and services. Jurisdiction to implement CALEA's provisions is shared by the Attorney General of the United States, who consults with state and local LEAs, and the FCC. Effective implementation of CALEA's provisions relies to a large extent on shared responsibility among these governmental agencies and the service providers and manufacturers subject to the law's requirements.

Great changes in technology have occurred over the past ten years, which have challenged the ability of LEAs to conduct lawful surveillance. Most notably, there has been a rapid shift from circuit-mode to packet-mode technologies, with an array of new services such as broadband Internet access and Voice over Internet Protocol (VoIP) now offered to consumers and businesses. The FCC has been proud to facilitate this communications revolution by minimally regulating these new services to promote increased competition and the introduction of new services for consumers and businesses.

These changes from a circuit-based world to a packet-based world will have a profound effect on the way we communicate. As my colleague Jeff Carlisle, now Chief of the FCC's Wireline Competition Bureau, noted just two months ago in testimony before this Subcommittee, voice is gradually becoming nothing more than one application of many over a multiuse digital network, where users may obtain a wide variety of services from multiple sources. For example, VoIP accelerates the migration to digital multiuse broadband infrastructures and internationalizes voice communications, allowing customers to buy voice applications from providers around the world. From the outset of this sea change, the Commission has stressed that important law enforcement obligations must be a part of any regulatory regime. And indeed, the very real threat of terrorism coupled with day-to-day criminal activity will not permit anything short of full CALEA compliance.

Against the backdrop of the advancing digital migration and facing these new challenges, the FCC is moving forward to ensure that CALEA's intent is fully carried out and that lawfully-authorized electronic surveillance is not compromised by new technologies—while at the same time not compromising the new technologies themselves.

PAST FCC RULEMAKINGS

In 1997, the FCC initiated a rulemaking proceeding to begin the implementation of CALEA, and over the next several years took a number of significant actions in that proceeding, focusing largely on circuit-switched technologies. Specifically, in an August 1999 *Second Report and Order,* the FCC concluded that the language and legislative history of CALEA provided sufficient guidance as to what the term "telecommunications carrier" means, such that the statute could be applied to particular carriers, their offerings and facilities. The Second Report and Order also stated that CALEA does not apply to certain entities and services, including information services and private network services. In a companion *Third Report and Order,* the FCC required that wireline, cellular, and broadband Personal Communications Services carriers implement all electronic surveillance capabilities of an industry-developed standard, as well as some additional capabilities requested by the Department of Justice and the Federal Bureau of Investigation.

CURRENT FCC RULEMAKING

In March 2004, the Department of Justice, Federal Bureau of Investigation, and Drug Enforcement Administration (collectively, Law Enforcement) filed a joint petition requesting that the FCC initiate a new rulemaking proceeding to resolve, on an expedited basis, issues associated with the implementation of CALEA. In its Petition, Law Enforcement maintains that outstanding implementation issues require immediate attention and resolution by the FCC, so that industry and federal, state, and local LEAs have clear guidance as CALEA implementation moves forward, particularly as communications technology changes. The Petition was placed on Public Notice on March 12, 2004; comments were due by April 12, 2004 and reply comments were due by April 27, 2004. The Commission received comments from LEAs, cable organizations, Internet and broadband companies/organizations, privacy and public interest groups, standards and technology groups, wireless companies/ organizations, and wireline companies/organizations.

On August 4, 2004, the FCC adopted a *Notice of Proposed Rule Making and Declaratory Ruling (Notice)* to launch a thorough examination of the appropriate legal and policy framework for implementing CALEA. In the item, the FCC states that it will be guided by several policy goals as it updates its CALEA policies: First, the

FCC wishes to ensure that LEAs have all of the resources that CALEA authorizes to combat crime and support Homeland Security. Second, the FCC recognizes that LEAs' needs must be balanced with the competing policies of avoiding impeding the development of new communications services and technologies and protecting customer privacy. Third, the FCC intends to remove to the extent possible any uncertainty that is impeding CALEA compliance, particularly for packet-mode technologies.

The *Notice* addresses a number of areas, including the applicability of CALEA to broadband Internet access and VoIP, capability requirements and solutions, compliance extensions, and cost and cost recovery issues. Each of these topics is discussed below.

Applicability of CALEA to Broadband Internet Access and VoIP

The *Notice* observes that CALEA applies to "telecommunications carriers" and exempts persons or entities insofar as they are engaged in providing "information services." The CALEA statute contains its own unique definition of the term "telecommunications carrier." Specifically, for purposes of CALEA, a "telecommunications carrier" is a person or entity engaged in the transmission or switching of wire or electronic communications as a common carrier for hire, but also includes entities that provide a replacement for a substantial portion of the local telephone exchange service if the FCC deems those entities to be "telecommunications carriers" as well. The *Notice* refers to this latter clause of the definition as the "Substantial Replacement Provision."

The *Notice* tentatively concludes that, where a service provider is found to fall within the Substantial Replacement Provision, it should be deemed a "telecommunications carrier" for purposes of CALEA, to which CALEA obligations would apply. If, at the same time, the FCC interpreted CALEA's information services exclusion to apply, it would present an irreconcilable tension; that is, particular service providers would find themselves at the same time subject to CALEA under the Substantial Replacement Provision and exempted from it by virtue of the information services exclusion. The *Notice* tentatively concludes that the better reading of the statute is to recognize and give full effect to CALEA's broader definition of "telecommunications carrier" and to interpret the statute to mean that where a service provider is determined to fall within the Substantial Replacement Provision, by definition it cannot be providing an information service for purposes of CALEA.

The *Notice* also tentatively concludes that facilities-based providers of any type of broadband Internet access, including but not limited to wireline, cable modem, satellite, wireless, and broadband access via powerline, are subject to CALEA because they provide replacement for a substantial portion of the local telephone exchange service used for dial-up Internet access service and such treatment is in the public interest. This tentative conclusion is based on the premise that broadband Internet access includes switching and transmission functionality and it replaces a substantial portion of the local exchange service used for narrowband Internet access.

The *Notice* observes that, at the time CALEA was enacted, Internet services were generally provided on a dial-up basis by two separate entities providing two different capabilities—a local exchange telephone company carrying the calls between an end user and its chosen Internet Service Provider (ISP), and the ISP providing e-mail, content, web hosting and other Internet services. In its *Report* on the CALEA statute, the House of Representatives was quite clear as to the status of these different entities under CALEA: The local exchange carrier providing the local exchange transmission service that enabled the call to that dial-up ISP—"the transmission of an E-mail message"—was covered as a telecommunications carrier providing a "plain old telephone service" or "POTS" functionality (a "phone call"). By contrast, the separate ISP was not subject to CALEA because the functions it provided—such as the storage of a message in an E-mail 'box' "—were "information services." The *Notice's* tentative conclusion respects the House's understanding and does not propose attaching CALEA obligations to services or applications that "ride over" the underlying broadband transmission, such as e-mail storage, web browsing capabilities and Internet gaming.

The *Notice* also tentatively concludes that providers of "managed" VoIP services, in which the provider acts as mediator to manage the communication between its end points and offers the service to the general public as a means of communicating with any telephone subscriber, including parties reachable only through the public switched telephone network (PSTN) are subject to CALEA. Such VoIP service providers offer an electronic communications switching or transmission service that replaces a substantial portion of local exchange service for their customers in a manner functionally the same as POTS service. The FCC believes that there is an overriding public interest in maintaining LEAs' ability to conduct wiretaps of on-going voice communications that are taking place over networks that are rapidly replacing the traditional circuit-switched network, yet providing consumers essentially the same calling capability that exists with legacy POTS service.

Further, the *Notice* observes that it appears that basic capabilities essential to LEAs' surveillance efforts, such as access to call management information (*e.g.*, call forwarding, conference call features such as party join and drop) and call set up information (*e.g.*, real time speed dialing information, post-dial digit extraction information) may not be reasonably available to the broadband access provider. Consequently, subjecting only the broadband access provider to CALEA without including managed VoIP service providers could undermine LEAs' surveillance efforts.

Capability requirements and solutions

The *Notice* seeks comment on telecommunications carriers' capability obligations under section 103 of CALEA. Section 103 requires telecommunications carriers to enable LEAs, pursuant to a court order or other lawful authorization, (1) to intercept, to the exclusion of other communications, wire and electronic communications carried by the carrier to or from a subject, and (2) to access call-identifying information that is reasonably available to the carrier, subject to certain conditions. Further, the interception of communications or access to call-identifying information is to be delivered to LEAs in a format that may be transmitted over the equipment, facilities or services procured by LEAs, to a location other than the provider's premises and in a way that protects the privacy and security of communications and information not authorized to be intercepted or accessed.

The *Notice* observes that CALEA defines call-identifying information as dialing or signaling information that identifies the origin, direction, destination, or termination of each communication generated or received by a subscriber by means of any equipment, facility, or service of a telecommunications carrier. The exact application of that term is not always clear in the context of broadband access and VoIP services. Call-identifying information may be found within several encapsulated layers of protocols, some of which may be considered packet content. The *Notice* invites comment as to how the FCC should apply that term for broadband and VoIP services. The *Notice* also invites comment on who may be in the best position to provide this information.

The *Notice* observes that telecommunications carriers may use whatever method they choose to satisfy CALEA's requirements. CALEA requires that LEAs and industry work cooperatively to develop standards that would serve as "safe harbors." In other words, if a telecommunications carrier employs an industry-developed standard, it would be deemed compliant with CALEA.

Under CALEA, any party may petition the FCC to address deficiencies in industry "safe-harbor" standards. While Law Enforcement has criticized certain of the industry standards, no petitions have been filed asking the FCC to intervene. The *Notice* invites comment as to whether standards for packet-mode technologies are deficient and thus preclude carriers from relying on them as safe harbors for complying with CALEA.

The *Notice* also invites comment on the feasibility of carriers relying on a trusted third party to manage their CALEA obligations. The trusted third party effectively acts as a surveillance service provider by collecting the packets from the carrier's network, extracting the information to which a LEA is entitled, and conveying it in an acceptable format to that LEA. Such an approach is already being used in both the United States and other parts of the world.

Compliance extensions

The *Notice* proposes several steps to ensure that telecommunications carriers comply with CALEA. CALEA section 107(c) provides that telecommunications carriers may request, and the FCC, after consultation with the Attorney General, may grant, extensions of time for CALEA compliance. The *Notice* proposes to restrict the availability of compliance extensions under CALEA section 107(c). The *Notice* also proposes to clarify the role and scope of CALEA section 109(b), under which carriers may be reimbursed by the Department of Justice for their CALEA compliance costs. The *Notice* specifies the information that would be required to be filed with Section 107(c) and 109(b) petitions. The Notice asks whether there are special concerns regarding small and rural carriers seeking additional compliance extensions, and, generally, proposes to afford all carriers with pending petitions a reasonable period of time (*e.g.*, 90 days) in which to comply with, or seek relief from, any determinations that the FCC eventually adopts in the rulemaking proceeding. Additionally, the *Notice* considers whether, in addition to the enforcement remedies through the courts available to LEAs under CALEA section 108, the FCC may take separate enforcement action against carriers that fail to comply with CALEA. The *Notice* tentatively finds that the FCC has general authority under the Communications Act to promulgate and enforce CALEA rules against carriers and non-common carriers.

Cost and cost recovery issues

In its Petition, Law Enforcement contends that CALEA places the financial burden of post-January 1, 1995 implementation on carriers and not LEAs. Law Enforcement requests that the FCC establish rules confirming that carriers bear the sole financial responsibility for post-January 1, 1995 CALEA implementation, unless otherwise specified by the FCC, recognizing that a specific carrier could have its costs reimbursed by the Department of Justice in the context of a CALEA section 109(b) petition. Related to this request, Law Enforcement asks the FCC to eliminate the issues of compliance costs as a basis for delayed compliance or non-compliance by establishing rules permitting carriers to recover CALEA implementation costs from their customers.

The *Notice* tentatively concludes that carriers are responsible for CALEA development and implementation costs for post-January 1, 1995 equipment and facilities. The *Notice* also seeks comment on cost recovery options that could reduce CALEA-related burdens otherwise imposed on carriers and their customers, particularly in rural areas. The *Notice* also asks for comment on how to assess the scope of CALEA-related costs in this proceeding. Commenters are requested to submit cost calculations and analysis, and to identify any conditions or factors that may affect the FCC's ability to determine the true scope of CALEA-related costs. The *Notice* refers to the Federal-State Separations Joint Board cost recovery issues for carriers subject to Title II of the Communications Act.

DECLARATORY RULING ON PUSH-TO-TALK SERVICES

The companion *Declaratory Ruling* grants Law Enforcement's request in the Petition and clarifies that commercial wireless "push-to-talk" services are subject to CALEA, regardless of the technologies that Commercial Mobile Radio Service providers choose to apply in offering them. In a prior decision, the FCC ruled that push-to-talk "dispatch" services that are interconnected to the PSTN are subject to CALEA. In effect, such push-to-talk service is a switched service that is functionally equivalent to a combination of speed dialing and conference calling. If push-to-talk "dispatch" service otherwise does not interconnect to the PSTN, the FCC found that it is not subject to CALEA.

Commercial mobile radio service providers are developing push-to-talk services based on use of packet technologies. Some parties asserted that such push-to-talk service is offered over a closed network and therefore should not be subject to CALEA. The *Declaratory Ruling* notes that CALEA is technology neutral; therefore, the choice of technology that a carrier makes when offering common carrier services does not change its obligations under CALEA.

CONCLUSION

As Chairman Powell noted in his statement on the CALEA Notice of Proposed Rulemaking and Declaratory Ruling: "[The Commission's] support for law enforcement is unwavering." As the Chairman also noted, the FCC's tentative conclusions in the *Notice* with respect to new packet-mode services such as VoIP is expressly limited to the requirements of the CALEA statute and does not indicate a willingness on the FCC's part to regulate those services as traditional telecommunications services. CALEA and other important social obligations can and will be continued without imparting upon carriers the full litany of analog, monopoly regulation. Similarly, the FCC is not proposing to regulate under the CALEA statute "non-managed" VoIP services, such as Instant Messaging, in which the service provider has minimal or no involvement in the flow of packets during the communication.

However, it is the FCC's unmistakable intent to ensure that LEAs have all of the electronic surveillance capabilities that CALEA authorizes to combat crime and terrorism and support Homeland Security. The FCC looks forward to developing a complete and comprehensive record before determining how best to proceed. The FCC will devote the necessary resources to expeditiously and responsibly complete this task.

The FCC is also cognizant that the Congress is currently contemplating legislation that may address CALEA. The FCC would welcome Congressional guidance in this area that would bring added certainty to the industry and lessen the risk of litigation. The Commission stands ready to provide whatever technical assistance that the Congress would find helpful in this regard.

I would like to thank you, Mr. Chairman, for the opportunity to appear before you today. This concludes my testimony and I would be pleased to answer any questions you or the other members may have.

Mr. UPTON. Thank you very much.

At this point we are going to take a brief adjournment as we have a series of votes on the floor. My sense is that we will be back about 12 o'clock. So we will take a 30 minute recess. We will come back at 12, and we will start with Mr. Baker when we come back.

Thank you.

[Brief recess.]

Mr. UPTON. We will resume.

I do not know. I guess the clock is still not working there at the table, but we'll resume. Is it working for them? There is a light on this side, and the light is out. So maybe that is the problem.

Mr. Baker, welcome.

STATEMENT OF STEWART A. BAKER

Mr. BAKER. Thank you, Mr. Chairman, members. I am Stewart Baker here on behalf of the Telecommunications Industry Association.

I am not here to suggest that wiretaps are not important or are not extraordinarily valuable for law enforcement. I used to be the General Counsel of the National Security Agency, and so I have some idea just how important it is to have good wiretap capability.

What I am here to suggest though is that saying that it is important for law enforcement to have wiretap capability is just the beginning of the inquiry. Preventing highway deaths is an important thing as well, and we could prevent highway deaths if we had a 30 mile per hour interstate speed limit. We have not done that even though preventing highway deaths is really important. The reason is because the costs are simply too high of implementing such a stringent regulation, and I think our concern is that looking over what the FCC has proposed, they proposed the equivalent of a 30 mile per hour speed limit on the ability of industry to innovate.

Ms. Parsky said all we want is for people to come in and consult with us, and the FCC has drafted a proposal that would create a vast regulatory machinery, enforcement machinery, that would enforce the requirement that people come in and consult.

But if you consult and you don't have an answer that the FBI likes, we know that the next step will be to go to enforcement, and so at the end of the day, this is a permission slip process. You need permission to innovate, and if you don't have the permission of the government when you want to roll out a new product, you can expect a law enforcement lawsuit and perhaps a cease and desist order.

That kind of tax on innovation is the biggest worry about the latest regulatory effort that law enforcement has launched here, and what I would suggest to the committee is that they take a look again at the way CALEA was written in the first place.

CALEA said we are going to set a standard, a performance standard. You have to be able to provide access to communications, and you have to provide reasonably available call identifying information. It is up to industry to figure out how to get there, and if they do not get there, then the Justice Department can take the company to court as soon as it can show that it has actually lost capability in an important case where they could not find some other way to get the information.

If they do that, then they will be able to impose penalties on the particular manufacturer and carrier that has not carried out its obligations. They have not brought any of those lawsuits, and instead they are proposing something that is a permission slip system.

In fact, I think if we just implemented CALEA as it is written, law enforcement would achieve its needs without imposing a tax on innovation.

I should say denying U.S. companies the ability to innovate without the permission of the Justice Department and the FBI does not mean innovation is going to stop. It just means it is going to happen someplace else, and in that regard, I would ask you to take a look at today's Wall Street Journal. The front page says, "China's telecom forays squeeze struggling rivals," and if you look at the chart that goes with that, you will see that the telecom manufacturers in China, the largest one there, 5 years ago was one-fiftieth the size of Lucent or Nortel. Today it is half the size, and the quotes suggest that what people are really worried about is what they will be doing 3 and 4 years from today in terms of their ability to penetrate this market.

They will develop products. They will test them. They will decide which ones are going to succeed in the market and which ones will not. They will do it in China. They will do it in Europe. They will do it in Southeast Asia.

And when they are ready, when they think, yes, this one will work, then they will bring those products to the United States, and they will sit down with the FCC and the FBI and the Justice Department and work out their CALEA obligations. U.S. companies though will not be able to do that unless they want to do their innovation abroad because they will not be able to try anything in a market until they have had the permission granted by the Justice Department and the FBI.

That strikes me as a fundamentally inappropriate way to approach this problem.

Thank you.

[The prepared statement of Stewart A. Baker follows:]

PREPARED STATEMENT OF STEWART A. BAKER ON BEHALF OF THE TELECOMMUNICATIONS INDUSTRY ASSOCIATION

Good morning. My name is Stewart Baker. Thank you for inviting me to testify today on behalf of the Telecommunications Industry Association (TIA). I am grateful for the opportunity to speak to you about the current status of law enforcement's ability to access new and ever-evolving communications systems, including broadband and Voice over Internet Protocol (VoIP) networks. TIA is a national trade association of 700 small, medium and large companies that provide communications and information technology products, materials, systems, distribution services, and professional services in the United States and around the world. In addition to representing its members on global policy matters, TIA is accredited by the American National Standards Institute, (ANSI), to develop American National Standards used by the industry. TIA also produces and co-owns SUPERCOMM, the largest annual communications industry conference and exhibition.

Let me begin by stressing that all of us on this panel want the government to have the tools that it needs to fight crime and terrorism. As a former General Counsel of the National Security Agency, I recognize that it is crucial to give law enforcement those tools. In fact, several months ago, I testified before the 9/11 Commission on the need for more aggressive use of government authorities to gather anti-terror information, and I cautioned about the risks of putting an undue emphasis on privacy concerns when pursuing terrorists. TIA also believes strongly that law enforcement needs to have the ability to conduct lawful surveillance of communications and to have lawful access to communications systems.

So we all can agree that ensuring lawful law enforcement access to evidence is an important goal—as important as preventing highway deaths or ensuring clean air or workplace safety. But if we've learned anything in the last twenty-five years of regulatory history, it's that we can't turn off our brains once we are told that a new regulation will serve an important social goal. No matter how important the goals they serve, some regulations make sense and some don't. Some go beyond statutory mandates. Some impose burdens that are nowhere near being cost-effective, stifling new industries and sending jobs overseas. This, unfortunately, is the kind of regulation that the Justice Department and the FBI support imposing today.

Of course law enforcement access is a good thing, at least when done within the law. But preventing highway deaths is also a good thing, and there's no doubt that we'd have fewer fatal accidents if the speed limit on interstate highways was lowered to 30 miles an hour. We won't do that, though, because the costs of such a regulation simply are not worth the added benefit. The same is true for wiretaps—except that today, there's a real risk that we will impose the wiretap equivalent of a 30 MPH speed limit on some of our most innovative and lucrative new industries.

The risk of over-regulating and stifling innovation is a risk that was well recognized ten years ago when Congress drafted the Communications Assistance for Law Enforcement Act (CALEA). I was in government when much of this drafting was done. CALEA was the result of a compromise that gave law enforcement a very carefully limited role in influencing the course of future technologies. Congress rejected the idea that the federal government should design or even have a veto over the design of new technologies. Instead, it set forth a very limited performance standard for wiretap access that would apply to a limited portion of the telecommunications industry. The law left lots of room for innovation and initiative. Industry was free to decide how to meet the wiretap requirement—industry had the right to set its own standards, which would provide a presumptively valid safe harbor for compliance, and individual companies that didn't like the standard remained free to try something else if they thought they had a better idea. Telecommunications technologies could be freely deployed without government interference, even if they did not have a perfect wiretap solution. Law enforcement could sue a carrier that deployed such technologies, but the carrier could defend itself by showing that full wiretap capability was not reasonably achievable in its system, or by showing that law enforcement could get the same information elsewhere.

TIA and its member companies rose to that challenge. TIA has led industry standards development efforts under CALEA, working jointly with the Alliance for Telecommunications Industry Solutions' Committee T1 to issue the leading CALEA compliance standard, J-STD-025, and the recent revision for packet-mode services, J-STD-025B. In fact, TIA member companies have gone well beyond what CALEA requires. For example, many companies that manufacture cable and Internet telephony hardware have already voluntarily built in intercept capabilities, despite uncertainty about whether CALEA applies to those services.

Despite this effort, disputes have arisen about what CALEA requires. Rather than continue to follow the dispute resolution processes established by Congress in CALEA, however, the Justice Department has asked the FCC to overturn key aspects of that carefully balanced statute. And in its proposed NPRM, the FCC seems ready to accept the Justice Department's invitation. The NPRM oversteps the Commission's regulatory authority in serious ways. First, the FCC proposes to write an entire new regulatory program to interpret and enforce CALEA, something that was not thought necessary when CALEA was enacted, or during the ten years thereafter. Second, the FCC seems willing to set aside CALEA's insight that industry knows more than government about how to design new telecommunications equipment. Rather than continue to encourage the development of common industry standards for giving law enforcement access to call information, the Commission seems poised to restrict the role of industry standards in CALEA. Third, the Commission is considering regulation that would cut off all avenues by which carriers can receive compensation for government-mandated network modifications—even going so far as to suggest that it may cut off reimbursements under a statute that the FCC has not interpreted, enforced, or administered for thirty-five years. Finally, TIA is concerned that the FCC will not allow adequate timelines for CALEA implementation.

On the first point, the FCC proposes that it should have a role in enforcing manufacturers' and providers' CALEA compliance, even though the statute clearly places enforcement in the hands of lawsuits to be brought by the Justice Department. But the FCC, citing its general enforcement authority under the Communications Act, tentatively concludes that it should promulgate CALEA rules that can be enforced against all entities deemed subject to CALEA.

The FCC's proposal is an end-run around the enforcement limits established in CALEA. Congress constructed a regime that gave carefully circumscribed enforcement power to the federal courts, and the Communication Act's general grant of authority to the FCC does not allow it to ignore the enforcement regime Congress established. In particular, the FCC's approach to implementing new enforcement regulations ignores the statutory defenses available to providers in enforcement actions. For example, in the enforcement regime established in CALEA, a company cannot be sanctioned unless law enforcement has no alternative method of getting the information it seeks through the enforcement action. Equally important, by threatening to use fines and cease-and-desist orders against noncompliant companies, the FCC will force innovators to get permission from the FCC and the Justice Department before deploying any new technology that falls into the wide grey zone created by the FCC's vague proposed regulations. An inventor who must get a government permission slip before trying out his invention is not likely to be first to market. While American innovators are still cooling their heels in Quantico, waiting to explain a new technology to the FBI Lab, their competitors in Singapore, China, Japan, and Europe will be manufacturing already. The U.S. market will end up a laggard, getting technologies after they've been sufficiently proven in the rest of the world to justify the engineering and lobbying costs needed to get an assurance of CALEA compliance.

At bottom, it is important that any enforcement framework allow for flexibility. Often, there is no simple answer to the question of how CALEA should be implemented. Instead, decisions in this area require a sophisticated balancing of the costs and benefits of various approaches. The CALEA framework is driven by industry standards and consultation between industry and law enforcement. And this negotiation-based approach is well-suited to the complex environment of CALEA compliance. To replace this framework with a top-down regulatory enforcement approach within the FCC would merely add another burdensome lawyer of regulatory pressure to an already complex CALEA-compliance process.

Second, TIA is concerned that in implementing its proposed CALEA rules, the FCC calls into question the sufficiency of the existing standards process, which has served as the backbone for industry compliance with CALEA. Industry-led standards development efforts are critical to the cost-effective and successful implementation of CALEA. Congress recognized the integral role of the standards process when it enacted CALEA. For example, when Congress had to make a choice between innovation and law enforcement control over CALEA compliance, Congress choose innovation, with its eyes wide open. Congress knew that the FBI wanted authority to oversee and even dictate the technical details of equipment manufacturers' CALEA-compliant solutions. But Congress rejected that approach, and instead enacted CALEA with a provision that prohibited law enforcement from requiring "any specific design of equipment, facilities, services, features, or system configurations." (47 U.S.C. § 1002(b)(1).)

At the same time, in Section 107(a) of CALEA, Congress explicitly noted the special role it gave to industry in creating standards to meet CALEA obligations. Section 107(a) "establishes a mechanism for implementation of the [CALEA] capability requirements that defers, in the first instance, to industry standards organizations." (H.R. Rep. No. 103-827, 1994 U.S.C.C.A.N. 3489, 3506 (1994).) But in order for this standards process to work effectively to address law enforcement's needs, industry needs to have the support of regulators. And right now, that support appears to be lacking. The FCC in its CALEA NPRM questions whether existing standards are deficient and whether it should only recognize standards produced by certain organizations.

Further, law enforcement has been uncomfortable with the fact that CALEA gives the lead standards role to industry. Since CALEA was enacted, law enforcement has wanted to guide, if not dictate, the detailed CALEA solutions that industry may implement. While this has created considerable tension between law enforcement agencies and industry throughout the standards process, there is no evidence to suggest that industry standards participants have acted in anything other than good faith.

In fact, TIA, its member companies, and other participants in TIA's standards activities have worked diligently—and cooperatively with law enforcement—for nearly a decade to adopt and improve CALEA standards and to ensure that law enforcement has access to appropriate, lawfully authorized electronic surveillance capabilities consistent with CALEA's statutory requirements. TIA's efforts led to the development of the J-STD-025 series of CALEA compliance standards, created at the expense of thousands of hours of industry experts' time and months of meetings.

Instead of scuttling the standards process altogether, law enforcement should be required to identify with specificity what aspects of what standards it is challenging, and the particular ways in which it deems the standards to be deficient. Industry should be given the opportunity to respond to law enforcement's concerns. Industry has demonstrated its responsiveness and diligence in developing standards in the past, and there is no reason to doubt that this level of cooperation will continue.

A leading role for industry in CALEA standards-setting is essential to further Congress's goal "to avoid impeding the development of new communications services and technologies." (H.R. Rep. No. 103-827, 1994 U.S.C.C.A.N. 3489, 3493 (1994).) Industry is by far best situated to design CALEA compliance standards in a complex, rapidly changing technology environment. An industry-led standards process permits U.S. companies to press forward with technological innovation, which is one of the key drivers of the U.S. economy in recent decades. At the same time, an industry-led standards process affords industry appropriate lawfully authorized electronic surveillance capabilities for evolving communications technologies.

The FCC also has suggested that perhaps CALEA standards should be set only by ANSI-accredited bodies. That is not what CALEA requires, and for good reason. TIA is an ANSI-accredited body, and it has written CALEA standards, so you might expect us to be comfortable with such a proposed limitation. But we are not. ANSI procedures call for consensus standard-making, and, in some instances, law enforcement has tried to use this requirement to defeat standards that all of industry has supported—by asking hundreds of sheriffs and local police to join the standards process at the last minute, for example, for the purpose of voting against the industry standard. In addition, an ANSI-accreditation requirement would encourage harsh tactics, such as the FBI's (now abandoned) effort to revoke TIA's accreditation after TIA adopted a standard that the FBI did not accept.

Third, TIA is concerned that manufacturers and service providers will be required to undertake expensive and burdensome network modifications in order to comply with CALEA under the FCC's proposed rules. Because the beneficiary of these changes are law enforcement agencies in the first instance and the general public in the last, one would expect that the cost of the changes would be carried largely by those parties. But the FCC's proposed rule puts the burden on industry, and it seems determined to make sure that there is no possibility of relief from the costs of CALEA. Instead, the FCC should reaffirm its previous conclusion that service providers may recover a reasonable share of CALEA costs that intercept law allows them to charge when carrying out a wiretap order. The principal mechanism for recovering those costs, Title III of the Omnibus Safe Streets and Crime Control Act of 1968, is far from the FCC's jurisdiction, and there is no need for the FCC to reach out now to determine that CALEA costs cannot be recovered under that statute.

Finally, TIA urges a reasonable timeline for requiring compliance with whatever rules the FCC eventually promulgates. Regulators and law enforcement must understand that industry needs a reasonable compliance deadline that creates enough space for equipment manufacturers, like the TIA members, to design and develop CALEA solutions well in advance of their actual deployment in the market.

In conclusion, I stress that, despite the crisis atmosphere fostered by the government, the Justice Department and law enforcement have never once used the enforcement powers that CALEA gives them. The only logical conclusion is that there has never been a single case—not one, not anywhere in the country, and not at any time in the last decade—in which the Justice Department thought it could prove that a carrier had failed to meet its CALEA obligation and that important evidence was being lost. Before throwing out CALEA as a failure and substituting a new FCC regulatory program that will slow innovation and saddle industry with heavy costs, we suggest that the government try using the tools that Congress provided ten years ago.

Mr. UPTON. Dr. Green.

STATEMENT OF RICHARD R. GREEN

Mr. GREEN. Thank you, Mr. Chairman and members of the committee.

I am Richard Green, President and CEO of Cable Television Laboratories.

This committee has been at the center of the technical revolution which has brought progressive enhancements to communication in the United States. It has been my privilege to testify before you on previous occasions on subjects related to emerging technology.

Today I appreciate the opportunity to testify on cable's leadership role in helping to facilitate law enforcement's legitimate access to voice over Internet protocol services. I speak to you today as a scientist who has devoted most of my professional career to the application of emerging technology.

In addition, Cable Labs conducts and funds research and development projects to help cable companies plan for the future and applies technologies to meet customers' needs. It is our purpose to foster and develop technologies which will support the United States in a leadership role and innovation.

Cable Labs was incorporated under the Cooperative Research Act. The act, which this committee played a key role in developing, encourages research and development among companies within an industry like the cable industry. I believe that we have been able to realize the potential of that act by, among other things, contributing to the development of a burgeoning broadband industry.

Turning to that issue which is before you today, the PacketCable project at Cable Labs has issued specifications, no worldwide standards, supporting among other services telephone services using advanced voice over Internet technologies. These specifications not only provide compliance with CALEA, but also introduce innovative Internet protocol technologies to insure that the United States remains a leader in the competitive marketplace of the future.

The cable industry has a history of cooperation with law enforcement. This was exemplified in the development of Cable Labs' PacketCable electronics surveillance specification developed during the period 1999 to 2004.

In 1999, at the request of cable operators and with the assistance of cable equipment manufacturers, Cable Labs published the first VoIP lawful electronic surveillance specification. This initiative was a volunteer effort by the cable industry to address requirements outlined in CALEA.

Law enforcement through the FBI and its contractors participated in the development of subsequent versions of that specification. These revisions reflect cable's willingness to work with law enforcement and to meet their needs even to the extent of adding additional capabilities and attendant costs to equipment.

The most recent version of the PacketCable electronic specification was published this year on July 23. Mr. Chairman, this development means that in spite of the numerous technological differences and complexities of VoIP, law enforcement will receive the same type of information and call content for voice services placed over PacketCable networks as in calls made with traditional wire line telephones.

Cable Labs has developed this technology not only to meet law enforcement's needs as addressed in CALEA, but also to address the public's privacy and security needs as mandated in the law. The devices and procedures in our specification are only activated pursuant to valid court orders and only gather information on the specific individual named in the court order.

We take great pride in a recent FBI press release commending the cable industry for its work in addressing the electronic surveillance requirement of Federal, state, and local law enforcement agencies.

In conclusion, Cable Labs and its member companies will continue our efforts to contribute innovative technologies to insure U.S. leadership in the world marketplace. We also look forward to continued cooperation with this subcommittee, the FCC, the FBI, the Department of Justice, and other Federal authorities in providing technical solutions to safeguarding our national security and the public's privacy and security needs.

Thank you very much, and I'll be pleased to answer any questions you might have.

[The prepared statement of Richard R. Green follows:]

PREPARED STATEMENT OF RICHARD R. GREEN, PRESIDENT AND CHIEF EXECUTIVE OFFICER, CABLE TELEVISION LABORATORIES, INC.

INTRODUCTION

Mr. Chairman, Mr. Markey, members of the subcommittee, I am Richard Green, President and CEO of Cable Television Laboratories, Inc. (CableLabs). It has been my privilege to testify before this subcommittee on previous occasions on subjects related to emerging technology. These topics have included High Definition Television in the 1980s, digital television, and broadband technologies in subsequent years. Today, I appreciate the opportunity to testify on cable's leadership role in helping law enforcement officials apply CALEA to modern digital telecommunications technologies. I am especially pleased to describe the industry's efforts—through CableLabs—to facilitate law enforcement's legitimate access to cable's Voice over Internet Protocol (VoIP) services.

I speak to you today as a scientist who has devoted a great deal of his professional career to questions involving the application of digital technology. The experience I gained during four years as Director of the CBS Advanced Television Technology Laboratory, five years as Senior Vice President of Operations and Engineering of PBS, and fifteen years as CEO of CableLabs gives me a special appreciation for the technical perspectives of manufacturers, cable operators, cable equipment manufacturers, and the need to be responsive to law enforcement's requests.

CABLELABS

CableLabs is a research and development consortium of cable television system operators serving North and South America. CableLabs conducts and funds research and development projects to help cable companies plan for the future and apply technology to meet consumers' needs. CableLabs was incorporated under the Cooperative Research Act. The Act, which this committee played a key role in developing, encourages research and development among companies within industries like the cable industry. I believe that we have been able to realize the potential of that Act by, among other things, contributing to the development of a burgeoning broadband industry, helping to spur the transition to digital TV, and facilitating the deployment of new digital services like VoIP.

For example, 29 million American homes now enjoy high-speed Internet access connections, and 18 million of those homes are served by cable's high-speed data service. The specifications for substantially all the cable modems used in those homes were developed at CableLabs. In the past, computer users knew that they could buy a modem that would work on any phone line. Cable industry leaders wanted their customers to be able to buy their own cable modem at retail and be confident that it would work on any cable system in North America. Through CableLabs' DOCSIS® (Data over Cable Service Interface Specification) project, that goal has been achieved. Cable's broadband service is providing an important new—and competitive—high-speed data highway into American homes.

The CableLabs process is open, cooperative, and as efficient as possible. We work to keep equipment development time to a minimum. We have pursued an approach similar to that used with cable modems to remove technical barriers for the deployment of telephone services over cable networks. The PacketCable project at CableLabs has issued specifications—now worldwide standards—supporting, among other services, telephone services using advanced voice over the Internet technologies. These standards go beyond compliance with CALEA but also introduce innovative Internet Protocol technologies to ensure that the United States remains a leader in the competitive marketplace of the future.

CABLE HAS COOPERATED WITH LAW ENFORCEMENT ON CALEA AND VOIP SINCE 1999

The cable industry has a history of providing law enforcement with the assistance it needs. This was exemplified in the development of CableLabs' PacketCable Electronic Surveillance Specification between 1999 and the summer of 2004. PacketCable is a cable network architecture that allows a cable operator to provide guaranteed-quality VoIP as well as other services such as video games. In 1999, CableLabs' PacketCable project, at the request of cable operators and with the assistance of cable equipment manufacturers, published the first VoIP lawful surveillance specification.

This specification was a voluntary effort by the cable industry to address CALEA in the event a cable operator's PacketCable service was deemed to be subject to CALEA. Law enforcement, through the FBI and its contractors, became involved in the development of subsequent versions of the PacketCable Electronic Surveillance Specification in 2001 with revisions to the specification published in 2003 and 2004. Each of these revisions reflects cable's willingness to work with law enforcement to meet law enforcement's needs—even to the extent of adding additional capabilities and attendant costs to equipment. The last version of the PacketCable Electronic Surveillance Specification was published on July 23, 2004, and provides solutions to *all* of the issues the FBI has identified with the previous versions of the specification. This now means that, in spite of technological differences and complexities, law enforcement will receive the same types of call identification and call content for calls placed over a PacketCable-compliant VoIP service as in calls made with traditional wireline telephones. (See Appendix I for a summary of the steps taken by CableLabs to develop the PacketCable specifications, 1999-2004.)

CableLabs developed its lawful surveillance specification not only to meet law enforcement's needs as are addressed in CALEA but also to meet obligations regarding the public's privacy and security needs as required by the law. CALEA expressly states that a telecommunications provider must ensure that subscriber privacy and security are protected for telecommunications and call-identifying information *not* authorized to be intercepted. The devices and procedures in CableLabs' specification are only activated pursuant to a valid court order and only gather information on the specific individual named in the court order.

The cable industry has met all of the FBI's needs with regard to VoIP. Specifically, CableLabs succeeded by July 2004 in resolving *every* issue on the FBI's "wish list" for CALEA compliance by cable's VoIP services, including:

- *Subject-initiated conference calls*—provides law enforcement with the content of subject-initiated conference calls.
- *Timing Information*—allows law enforcement to correlate call identifying information with call content.
- *Subject-initiated dialing and signaling*—provides law enforcement with access to all subject dialing and signaling information such as use of flash hook (call waiting) and feature keys.
- *In band/out-of-band signaling*—notifies law enforcement whenever subject's service sends a tone or other network message such as if a line is ringing or busy.
- *Party Hold/Join/Drop*—allows law enforcement to identify the active parties to a subject-initiated call.
- *Dialed Digit Extraction*—provides law enforcement those digits dialed by a subject during a call.

Testing of cable equipment built to these specifications will begin in February 2005, and products that do not meet the latest version of the PacketCable Electronic Surveillance Specification will not be CableLabs' certified—nor are they likely to be purchased by cable operators.

The success of the PacketCable Electronic Surveillance Specification is demonstrated in: (1) its being the only VoIP CALEA "Safe Harbor" specification listed on the FBI's AskCALEA website; (2) its consideration by other VoIP-related organizations; (3) the FCC's public commentary noting cable's contribution to the lawful electronic surveillance of VoIP calls; and (4) the FBI's cooperation in, and contribution to, the specification's development and its subsequent positive comments on the specification and the CableLabs' process.

The cable industry, through CableLabs, continues to provide technical assistance to law enforcement and has worked with the FBI on how law enforcement may collect the information it lawfully needs from subjects using PacketCable-based VoIP services. We take great pride in comments from a recent FBI press release in which Kerry Haynes, FBI Assistant Director for the Investigative Technology Division, states:

> [The latest version of the PacketCable Electronic Surveillance Specification] is an extremely positive development for the cable industry that ultimately will empower federal, state and local law enforcement agencies with the technical capability to continue to protect the public by effectuating court-authorized electronic surveillance investigations. We look forward to working with the industry in its development of technical solutions based on this standard and with companies as they implement those solutions into their IP networks.

CABLE WAS THE FIRST BROADBAND PROVIDER IN 2004 TO COOPERATE WITH THE FEDERAL COMMUNICATIONS COMMISSION ON APPLYING CALEA TO VOIP SERVICES

The cable industry has long recognized that certain IP telephony services may become a replacement for some of the uses of traditional telephony, and that—at some point—providers of such services could reasonably be expected to provide efficient and effective means to allow law enforcement access to telecommunications over such services.[1] For this reason, the cable industry, led by CableLabs and the member companies of NCTA, has voluntarily sought to comply with the substance of CALEA's requirements in developing its PacketCable architecture for VoIP. In particular, as I mentioned above, the industry has devoted substantial resources to developing several PacketCable Electronic Surveillance Specifications for use as "safe harbors" under 47 U.S.C. § 1006(a)(2).[2]

In 2004, the Department of Justice (DOJ) and Federal Bureau of Investigation (FBI) asked the Federal Communications Commission (FCC) to issue a declaratory ruling determining immediately—*i.e.*, without awaiting the outcome of the Commission's rulemaking on IP-Enabled Services[3]—that CALEA applies to various kinds of IP telephony ("Broadband Telephony") as well as to cable modem service and other forms of high-speed Internet access ("Broadband Access"). *In their submissions to the FCC, most communications industries urged the Commission to reject the Administration's requests—except for the cable industry.*

Cable companies—through their trade association, NCTA—took a different position. They supported the issuance of a declaratory ruling by the FCC that providers of Broadband Telephony are properly viewed as "telecommunications carriers" for purposes of CALEA, subject to two qualifications. First, the FCC should include within the scope of its ruling all similarly-situated providers of Broadband Telephony, including services like Vonage and AT&T's CallVantage. Second, the Commission should make clear that, when services like Vonage and AT&T's CallVantage are provided over the facilities of cable operators or other companies, the responsibility for complying with CALEA lies with the Broadband Telephony provider, not the facilities owner.

In addition, NCTA supported the issuance of a Notice of Proposed Rulemaking (NPRM) addressing whether Broadband Access should be made subject to CALEA in due course. The ultimate decision on the merits here, however, raises more complex issues. Until now, there has never been substantial reason to expect that cable modem service might ever be subjected to CALEA. Thus, there has been little investigation or debate concerning the public policy and law enforcement objectives in developing of CALEA-related technical requirements for the equipment that cable operators use to provide the service. However, the cable industry and CableLabs will continue to work with the United States Government to ensure that law enforcement is able to access lawfully the information needed to safeguard our national security.

In response to the joint DOJ/FBI petition, the FCC recently commenced a rulemaking on CALEA compliance issues. It tentatively concluded that most VoIP services would be subject to CALEA—essentially echoing the cable industry's legal rationale. It also tentatively concluded that Broadband Access should be subject to

---

[1] *See* H.R. Rep. No. 103-827, at 9 (1994) ("House Report") (purpose of CALEA is "to preserve the government's ability. . . to intercept communications involving advanced technologies").

[2] The most recent version of the PacketCable Electronic Surveillance Specification is available at http://www.packetcable.com/downloads/specs/PKT-SP-ESP-I03-040113.pdf. Prior versions, which provide safe-harbor protection to providers that have already installed equipment that is compliant with those versions, are available at http://www.cablelabs.com/specifications/archives/

[3] *IP-Enabled Services,* Notice of Proposed Rulemaking, WC Docket No. 04-36, FCC 04-28 (rel. Mar. 10, 2004) ("IP-Enabled Services NPRM").

CALEA. Cable companies and the NCTA will submit their own individual comments on the specifics of such a proposal to the FCC.

CONCLUSION

CableLabs and its member companies—who also belong to NCTA—look forward to continued cooperation with this subcommittee and other Federal authorities in safeguarding our national security. I would be pleased to answer any questions you might have.

APPENDIX I

SUMMARY OF THE DEVELOPMENT OF CABLELABS' LAWFUL SURVEILLANCE ARCHITECTURE WITHIN THE PACKETCABLE SPECIFICATION FOR VOIP

I. First Version Published December 29, 1999
  a. Drafted at CableLabs members' request in the event VoIP (using cable's PacketCable architecture) was deemed to be subject to CALEA.
  b. Developed by MSOs, cable equipment manufacturers pooling their intellectual property and MSO legal community.
  c. Established basic surveillance needs:
    i. Demarcation point between MSO and law enforcement (delivery of call content and call identification from the MSO to law enforcement).
    ii. Delivery Function (in which the copied packets are delivered to law enforcement's Collection Function).
    iii. Intercept Access Points within the PacketCable Architecture.
    iv. Basic capabilities for delivery of VoIP call information and VoIP call content to law enforcement.

II. Second Version Published August 15, 2003
  a. June 2001—The FBI became involved and submitted engineering changes to the PacketCable Lawful Surveillance Specification.
  b. November 2001—CableLabs forms a focus team of cable equipment manufacturers to address the FBI's requested changes and resolve technical issues with the first version of the specification.
  c. New capabilities added:
    i. Law enforcement receives information on subject initiated signaling (signals such as number dialed, flash hook, feature keys).
    ii. Law enforcement receives information on network initiated signaling (such as call connection and hang up).

III. Third Version Published January 13, 2004
  a. Addressed additional FBI engineering change requests submitted just prior to the publication of the second version of the specification.
  b. Addressed minor technical issues within the second version of the specification.
  c. Coordinated specifications information for the Delivery Function—Collection Function Interface.
  d. New capabilities added:
    i. Report of IP specific "call data" to law enforcement for trap and trace and pen register.
    ii. Three-way calling information.
    iii. All relevant punch list items met save some conference call information and dialed digit extraction (collection of numbers called after call was initiated, such as PIN numbers).

IV. Fourth Version Published July 23, 2004
  a. MSO push to support solutions to collection of all FBI requested conference call information and collection of digits dialed after call is initiated (dialed digit extraction).
  b. FBI provided a list of specific comments to the current specification in May, 2004.
  c. Dialed Digit Extraction solution reached by adding additional capabilities, at additional cable operator cost, to the Delivery Function.
  d. New Capabilities:
    i. Dialed Digit Extraction.
    ii. Party Hold/Join/Drop (knowing when someone joins a conference call, leaves a conference call or goes on hold).
    iii. Transcoding between the Delivery Function and the Collection Function.
      1. Translation of the many codecs (means of translating packets into voice) that cable operators use or may be used to just a few codecs.

2. Lessens the number of codecs law enforcement needs to support (helpful for rural law enforcement with small budgets).
3. FBI originally wanted to limit the number of codecs cable operators may use. This CableLabs' solution allows for future growth and technological change while meeting law enforcement's needs.

e. Punch List Items are now all addressed:
  i. *Subject-initiated conference calls*—provides law enforcement with the content of subject-initiated conference calls.
  ii. *Timing Information*—allows law enforcement to correlate call identifying information with call content.
  iii. *Subject-initiated dialing and signaling*—provides law enforcement with access to all subject dialing and signaling information such as use of flash hook (call waiting) and feature keys.
  iv. *In band/out-of-band signaling*—notifies law enforcement whenever subject's service sends a tone or other network message such as if a line is ringing or busy.
  v. *Party Hold/Join/Drop*—allows law enforcement to identify the active parties to a subject-initiated call.
  vi. *Dialed Digit Extraction*—provides law enforcement those digits dialed by a subject during a call.

f. Testing of equipment to begin February 2005:
  i. Products that do not meet the latest version of the PacketCable Electronic Surveillance Specification will not be CableLabs certified.
  ii. Cable operators prefer to buy CableLabs-certified equipment.

Mr. UPTON. Thank you.
Mr. Dempsey.

STATEMENT OF JAMES X. DEMPSEY

Mr. DEMPSEY. Mr. Chairman, members of the subcommittee, good afternoon. Thank you for this opportunity to testify.

Mr. Chairman, nobody denies the interests of the government in being able to intercept terrorist communications. These are obviously extremely important. Let us even assume that they are paramount. Let us assume that they trump all other public policy interests. Forget about privacy; forget about cost, innovation, competition, network security.

Even if we assume that law enforcement and national security interests are the only interests at stake, if you look at the record, you would have to conclude that CALEA is not the right statute for addressing law enforcement interests in accessing the Internet and that the FBI is not the right agency to be regulating the design of information and services and Internet access.

Now, I am happy to discuss at length the language and intent of CALEA. Congress was as clear as it could be in CALEA that it did not apply to the Internet, that it was intended for the circuit switched world of the PSTN. Congress used not only a belt in saying that CALEA applied only to telecommunications common carriers, but it used suspenders as well and said that information services and Internet services were exempt from CALEA.

It then used some safety pins and said that even if Internet services became a replacement for a substantial portion of the traditional telephone network, it was still excluded from CALEA.

The FCC issued an NPRM last month and was so results oriented because it was so focused on this compelling interest of fighting terrorism that it decided to ignore the statutory language. Three of the five Commissioners filed separate statements indicating doubts about whether the Commission's rationale would withstand scrutiny.

But let's leave aside even the question of statutory interpretation. Let's focus on the record of CALEA implementation in the plain, old telephone network, which was supposed to be the easy part.

The Department of Justice's own Inspector General found in a report issued in April this year, "deployment of CALEA technical solutions for electronic surveillance remains significantly delayed. Most of the authorized funds have been depleted. Even by the FBI's own estimate, hundreds of millions of dollars more are needed."

Most troubling, the IG said, "CALEA compliance software has been activated on only 10 to 20 percent of wire line equipment," and the IG found the FBI was unable to demonstrate the extent to which lawful electronic surveillance has been adversely impacted by the lack of CALEA implementation. In other words, they could not show whether this delay in enforcement or implementation of CALEA made any differences.

What went wrong? CALEA, when it was enacted in 1994, was filled with checks and balances. It was a very nuanced statute, but it has become a straightjacket. The way it has been interpreted by the FCC, it has given the FBI the ability to design and dictate very specific capabilities, very specific features to be built into the public switched network in ways that Congress never contemplated.

Stewart Baker is the one who coined the phrase "because of CALEA the FBI has become a telecom regulatory agency." FBI Director Freeh in 1994 came before Congress and said that CALEA was not intended to create a location tracking capability for cell phones. As soon as that legislation was passed, the FBI came to the industry and said, "We want you to build in a location tracking capability." And they got it. The FCC gave it to them.

The Director said we only want to get dialed number information on pen registers. After the legislation was enacted, the FBI came back and said, "No, we want to know every time a party goes on hold. We want to know whether the phone rang or had a busy signal." The FCC ordered that those features be built in.

The FBI said they only wanted to preserve their traditional surveillance capabilities. After the legislation was enacted, they came back and said, "We want to have the ability on a conference call to identify every separate party," even though they admitted that that was not a capability that they had had in the traditional telecom environment.

Now, how do we go forward? The first step has to be to create a factual record, to identify what are the specific problems. The 101 page NPRM of the FCC has absolutely no factual discussion of what are the problems. The FBI's petition has three conclusory sentences. Somebody needs, and I think it has to be on the public record; I think this committee has a role in it; needs to dig in on those facts and find out what the problems are.

Second, the solution has to be consistent with the decentralized and innovative architecture of the Internet. There may be some very simple solutions out there. The so-called trusted third party model has been put forward. There are actually companies now who have the ability to analyze Internet communications and deliver them to law enforcement. Why can't law enforcement simply acquire that capability itself?

Mr. Chairman, you alluded in your opening statement to partnership. CALEA was intended as a partnership. It has not worked out that way. Here we have the cable industry currently not subject to CALEA out there developing a standard and cooperating with industry. I think there is a pretty good bet that if they were brought under the requirements of CALEA, the FBI would find something wrong with that standard and would ask for even more and would constantly go back.

That is what delayed implementation of CALEA. The FBI could have had 90 percent of the capability they were seeking 4 or 5 years ago, but instead they kept driving for this 100 percent concept every little additional piece, and we are left now with CALEA not even fully implemented for the traditional telephone network.

My organization, the Center for Democracy and Technology, is happy to work with the committee to work through these issues, to try to build this factual record, to try to drill down and to develop solutions that are appropriate to the Internet.

Thank you, Mr. Chairman.

[The prepared statement of James X. Dempsey follows:]

PREPARED STATEMENT OF JAMES X. DEMPSEY, EXECUTIVE DIRECTOR, CENTER FOR DEMOCRACY & TECHNOLOGY [1]

Chairman Upton, Congressman Markey, and Members of the Subcommittee, thank you for the opportunity to testify today.

Especially in the face of terrorism, the question of law enforcement access to communications systems is vitally important. However, the Justice Department and the Federal Communications Commission are trying to force the Internet into a 20th century mold. In terms of innovation, cost, privacy, network security, *and* national security, this is the wrong approach. Instead of making the Internet look like the telephone system of the past, the FBI and other law enforcement agencies need to acquire in-house capabilities to analyze digital communications. They should use the Internet, not try to control it. Keeping pace with technology should not require slowing it down.

Law Enforcement Mandates Designed for the Telephone Network Are Not Suited—Nor Are They Needed—for the Internet

To understand why the Justice Department's approach is unnecessary, unwise and unlikely to be effective, think of the ways in which the Internet is different from the traditional telephone network of the past. In the old days, when law enforcement agencies first started lawfully wiretapping telephones, the Ma Bell monopoly owned and controlled the entire network, right down to the phone on your desk. Such a centralized system was reliable, but it was limited. Innovation was discouraged. Competition was essentially non-existent. Prices were regulated but relatively high, and usage was cautious.

Now consider the Internet. It is open, competitive, decentralized. It supports a multiplicity of applications, not only voice, but also photography, data, and video. It supports one to one, many to one, and one to many communication. It pushes control to the edges, giving users far more choices then they ever had. It has no gatekeepers. It intermeshes wireline, wireless, cable and satellite. It is innovative, inexpensive, and global. Education, commerce, medicine and government have reaped the benefits.

In the context of today's hearing, the Department of Justice complains about the Internet's diversity, but in many ways the digital age is the age of surveillance.

[1] The Center for Democracy and Technology is a non-profit, public interest organization dedicated to promoting civil liberties and democratic values for the new digital communications media. Our core goals include enhancing privacy protections and preserving the open architecture of the Internet. Among other activities, CDT coordinates the Digital Privacy and Security Working Group (DPSWG), a forum for computer, communications, and public interest organizations, companies and associations interested in communications privacy and security issues.

More personal information than ever before is transmitted, collected and stored in electronic form. In many ways, law enforcement has embraced the digital revolution. Every year, the number of wiretaps goes up. Undercover agents lurk in Internet chats. Police track suspects through cell phones and reconstruct past movements from EZ Pass logs. The FBI can plant on your computer a keystroke monitor to copy letters you never send. Agents seize computer disks holding information that would fill truckloads if printed out. Voluminous dialing records are analyzed by computer. Conversations intercepted in New York are shipped across country for translation. A computer in Russia can be searched from the US.

So despite some of the dire rhetoric you may hear, the Internet is already tappable today, both legally and from a technical standpoint. The government has full legal authority to tap broadband Internet access and Internet communications of all kinds. The government also has all the legal authority it needs to compel broadband access providers and Voice over Internet Protocol ("VoIP") service providers to co-operate with court orders for interception. 18 U.S.C. § 2518(4). And from a practical standpoint, law enforcement agencies currently have and in the foreseeable future will continue to have the capability to intercept communications over broadband. In some ways, interception may be less convenient, in that law enforcement may have to go to different entities to obtain content and routing information. And given the diversity of services, the information will come in different formats and law enforcement will have to work harder to determine what it is intercepting. In other ways, however, Internet surveillance will be easier, in that the digital nature of communications makes them easier to analyze, store, and retrieve. Last year, for example, according to the government's official Wiretap Report, out of 1,442 authorized wiretaps nationwide, the "most active" was the interception of a broadband Internet line.[2]

The only question—and it is a big question—is whether additional authority is needed for the government to insert certain features into Internet services to make them easier to tap. Answering that question requires, first, a detailed, technical inquiry into whether there are any problems associated with Internet surveillance. It then requires a detailed, technical exploration of how those problems can be solved, with consideration of the various costs and risks of different solutions. Throughout, it is important to keep in mind the ways in which the architecture of the Internet is different from the traditional telephone network.

CALEA Was Designed for the Traditional Public Switched Telephone Network

In the 1990s, Congress conducted such an inquiry with respect to the public switched telephone network. It found that there were some problems posed by then relatively new technology in the PSTN, and it concluded that the solutions lay in redesign of the central office switches of the telephone companies. The result was the Communications Assistance for Law Enforcement Act ("CALEA").

CALEA is a 20th century statute for 20th century technology. CALEA was designed for the centralized, relatively monopolized, and circuit switched world of the traditional telephone common carriage—entities already subject to a range of regulatory burdens. The proposed solution focused on central office switches. That is where the documented problems were. The carriers operating those switches used for routing and billing purposes the information they thought the government wanted. The switch manufacturers thought it would be relatively easy to build in the ability to meet the government's requests as they were described in the legislative hearings.

CALEA has not worked all that well even for the PSTN—the government ended up demanding a lot more functionality, including features not available with the traditional wiretaps—but the Internet is fundamentally different from the PSTN and requires a different approach.

Congress was crystal clear—CALEA was not intended for the Internet. To make this point, Congress took not merely a belt and suspenders approach, but added safety pins as well. It said that CALEA applied only to common carriers, and only to the extent that they are providing telecommunications services. It excluded information services, and it said that even if an information service became a substantial replacement for the PSTN in a particular region, it would still be excluded from the requirements of CALEA.

At the time, the term "information services" was shorthand for the Internet and the applications running over it (among other services). The term "information services" was broadly defined to cover current and future advanced software and soft-

[2] "Report of the Director of the Administrative Office of the United States Courts on Applications for Orders Authorizing or Approving the Interception of Wire, Oral, or Electronic Communications," issued April 30, 2004, available at http://www.uscourts.gov/wiretap03/contents.html.

ware-based electronic messaging services, including email, text, voice and video services. Narrowband Internet access and Internet applications like email fit squarely within the definition. As the broadband Internet has evolved, it continues to be outside the scope of telecommunications common carriage, and Internet-based telephony services, like all other Internet applications, fit squarely within the definition of information services.

The legislative history confirms the plain meaning of the statute. The Committee Report states that CALEA obligations "do not apply to information services, such as electronic mail services, or on-line services, such as CompuServe, Prodigy, America On-line or Mead Data, or Internet service providers." *Telecommunications Carrier Assistance to the Government,* H.R. Rep. 103-827(I), at 23 (Oct. 4, 1994) ("House Report"). As the FBI Director testified, CALEA was "narrowly focused on where the vast majority of our problems exist—the networks of common carriers, a segment of the industry which historically has been subject to regulation." [3]

Reading the statute and legislative history, both the FCC itself and the D.C. Circuit in the past held that CALEA does not apply to the Internet. In 1999, the FCC concluded that information services "such as electronic mail providers and on-line service providers" are exempt from CALEA. *In the Matter of Communications Assistance for Law Enforcement Act,* Second Report and Order, 15 FCC Rcd 7105, at ¶26 (1999). The D.C Court of Appeals stated, "CALEA does not cover 'information services' such as email and internet access." *United States Telecom Ass'n v. FCC,* 227 F.3d 450, 455 (D.C. Cir. 2000).

The FCC has recently issued a Notice of Proposed Rulemaking, tentatively concluding that CALEA should apply to broadband Internet access and "managed" Voice over Internet Protocol ("VoIP") services. The NPRM is purely results-oriented. The Commission looked at the urgency of the terrorist threat, and jumped straight to the conclusion that CALEA should be extended to the Internet. To do so, it admitted that it was ignoring the language of the Act and contradicting its own earlier decisions about the regulatory status of broadband access. Three Commissioners hinted in separate statements that the Commission's rationale would not withstand judicial scrutiny.

Congress Needs to Conduct a Factual Inquiry

The first step in responding to the arguments of the Department of Justice must be a clear showing of need: what are the problems that law enforcement is encountering? In the early 1990s, during the George H.W. Bush Administration and then in the Clinton Administration, when the FBI began complaining that technological changes in the PSTN were interfering with law enforcement's ability to carry out wiretaps, Congress refused to adopt a sweeping regulatory mandate. Instead, Congress insisted first and foremost on a factual inquiry into what exactly were the problems being encountered by law enforcement. Hearings were held. The General Accounting Office conducted two studies. The FBI surveyed its field offices twice. Industry and law enforcement convened action teams to study the concerns of law enforcement and possible solutions. At the end of the process, industry representatives agreed that new technologies were defeating law enforcement surveillance. Some of the problems had to do with features such as call forwarding and speed dialing. Others had to do with the transition to multiplexed lines and fiber optic cables. Most had to do with the lack of sufficient capacity on switches to simultaneously accommodate a large number of intercepts.[4]

In 2004, the DOJ/FBI petition and the FCC's 101 page NPRM are devoid of any factual discussion of problems justifying extension of CALEA. In the 1990s, when arguing for CALEA, the FBI Director talked about a de facto repeal of the wiretap laws. The lack of capacity to accommodate multiple intercepts on wireless switches, which accounted for the majority of problems documented in the 1990s, represented a complete shutout for law enforcement. But in the Internet context, the FCC's recent NPRM refers to problems such not getting exactly the same information on broadband communications that is available in the PSTN, or not having the information delivered in a familiar format. These are not the magnitude of problem that justified Congress adopting CALEA for the already well-regulated telecommunications common carriers—they surely do not justify a regulatory mandate for the Internet. Is there a problem of not having access at a single point to all features

---

[3] Testimony of Louis Freeh before the Joint Hearing of the Technology and Law Subcommittee of the Senate Judiciary Committee and the Civil and Constitutional Rights Subcommittee of the House Judiciary Committee, Mar. 18, 1994, available at http://www.eff.org/Privacy/Surveillance/CALEA/freeh_031894_hearing.testimony.

[4] *Telecommunications Carrier Assistance to the Government,* H.R. Rep. 103-827(I) at 14-16 (Oct. 4, 1994).

and services used by a surveillance target? Even with respect to the PSTN, CALEA was not intended to guarantee one-stop shopping for law enforcement. Are there difficulties in determining which service provider or which kinds of services a particular suspect is using? If so, that seems to be an unavoidable byproduct of the diversity of services that our telecommunications policy has wisely fostered, not a problem requiring design mandates.

The second step should be a showing of what would a design mandate for the Internet look like. In this regard, Congress would have to be very careful and insist on more specificity than it did in 1994. In applying CALEA to the PSTN, the FCC adopted an elastic interpretation of CALEA's definitions, requiring carriers to build into their systems surveillance features that went beyond what had been available to law enforcement in traditional systems. For example, the FCC gave five different meanings to the word "origin" in the definition of "call-identifying information."[5] Such flexibility applied to the Internet could produce endless demands.

In some ways, the debate today is reminiscent of the encryption debate of 10 years ago. Law enforcement agencies felt threatened by encryption. They thought it meant terrorists and drug dealers could communicate in perfect confidentiality. The government argued that encryption had to be "dumbed down" or built with backdoors for easy government access. After a long debate, Congress and the Administration decided that the technology should not be controlled. Law enforcement and intelligence agencies adjusted. Beginning with the 2000 Wiretap Report, the government has been required to report on whether encryption was preventing law enforcement officials from obtaining the plain text of communications intercepted pursuant to the court orders. So far, the government has not reported a single wiretap frustrated by encryption. In 2003, no federal agencies conducting wiretaps reported that encryption was encountered. For state and local jurisdictions, encryption was reported to have been encountered in one wiretap in 2003; however, the encryption was not reported to have prevented law enforcement officials from obtaining the plain text of communications intercepted.

CALEA Has Not Been Very Successful Even as Applied to the PSTN

Even as applied to the relatively centralized PSTN, CALEA has not worked well. The FBI and DOJ admitted as much in their petition to the FCC. Indeed, their petition was almost schizophrenic: the first half argued that the Internet should be brought within the regulatory scheme of CALEA while the second half laid out a litany of delays, confusion and controversy under CALEA as applied to the PSTN.[6] The DOJ and FBI stated that the CALEA implementation process "is not working." Petition, at 38. They cited "problems and delays," *id.* at 53; a "seemingly endless cycle of extensions that have consistently plagued the CALEA compliance process," *id.* at 55; and more "problems and delays," *id.*

This record of disfunctionality is confirmed by a report by the Office of the Inspector General (OIG) of the U.S. Department of Justice, issued on April 7, 2004. The OIG's biannual audit, mandated by CALEA, evaluates the progress of CALEA compliance, and finds broad problems. The report notes that costs of CALEA for the PSTN have been much higher than Congress anticipated. "Most troubling, according to FBI estimates, CALEA compliant software has been activated on only 10 to 20 percent of wireline equipment." The report also shows that the FBI's insistence on it "punchlist" has caused enormous problems within the CALEA standards setting efforts of industry. Most remarkably, the report finds that the FBI "was unable to demonstrate the extent to which lawful surveillance has been adversely impacted by the lack of CALEA implementation."[7]

Simply put, CALEA has proven to be a flawed statute. As to why, there is probably enough blame to go around. One key factor is that, contrary to Congress' intent, the FBI exercised de facto power to impose specific design mandates on the PSTN, and it used this power to impose on industry surveillance features that not only went beyond the capabilities of the traditional telephone system but that could have been procured by law enforcement itself for less expense. For example, the

[5] "Origin" refers, of course, to the phone number of the party initiating a call. The FCC ruled, however, that "origin" also means the signal indicating that a call is waiting, Third Report and Order, *In the Matter of Communications Assistance for Law Enforcement Act,* 14 FCC Rcd 16794 (1999) ¶82; use of the flash key on the telephone to switch back and forth between two established calls, *id.*; putting a party on hold, *id.* ¶74; and the location of a wireless phone caller at the beginning and end of a call, *id.* ¶44.

[6] Joint Petition for Rulemaking to Resolve Various Outstanding Issues Concerning the Implementation of the Communications Assistance for Law Enforcement Act, FCC RM-10865 (filed Mar. 10, 2004).

[7] "Implementation of the Communications Assistance for Law Enforcement Act by the Federal Bureau of Investigation," available at http://www.usdoj.gov/oig/audit/FBI/0419/final.pdf.

FCC imposed at least $120 million in costs on industry to obtain one feature known as "dialed digit extraction," which requires local exchange carriers, after call set-up, to reach into the content of the communications and extract additional dialed numbers, such as the numbers called on a long distance calling card. The FBI could have obtained the information it wanted by going to the providers of long distance services, but it wanted to obtain the information more conveniently through the local phone system. Indeed, the FBI could have purchased the extraction devices itself and attached them as necessary, a solution that the FBI itself estimated would cost no more than $20 million a year, but instead the FBI insisted that all carriers install them on all switches.

Going Forward: Meeting Law Enforcement Needs in a Way Suited to the Decentralized, Innovative Internet

Clearly, a different approach is needed for the Internet. As we suggested at the outset, that solution must take into account the decentralized, innovative, user-controlled nature of the Internet.

There are three possible approaches: One is the internal approach of CALEA, which DOJ is proposing to impose on the Internet, requiring extensive standards processes, detailed specifications, and FCC enforcement to require access providers and service providers to build capabilities into their equipment and software. The second is what the FCC refers to as the "trusted third party" approach, in which a service bureau sits between the service provider and the law enforcement agency, analyzing packets, extracting signaling information, and formatting it for the convenience of law enforcement.

There is a third approach, which is suggested by the service bureau model: Instead of forcing industry to redesign its products and services to meet government specifications, law enforcement should itself develop (or acquire from the service bureaus) the capabilities to analyze packet communications. In other words, law enforcement should develop the capability to extract call-identifying information from packet streams. Even CALEA only requires carriers to deliver call-identifying information to law enforcement—it imposes no formatting requirements on service providers. Moreover, the government will have to develop the capability to analyze packets in-house anyhow, because it will have to be able to deal with sophisticated criminals who can entirely avoid service providers and communicate directly and with custom-built protocols. Perhaps Congress should appropriate additional funds to the FBI to keep pace with technology in this way and to support state and local law enforcement efforts to do the same.

This third approach—a fundamentally non-regulatory approach—illustrates how the assumptions that applied to CALEA in the PSTN are probably inapplicable to the Internet. The Internet may not need a detailed technical standard the way the circuit switched environment does. The call processing technology that once existed solely in the control of the monopolistic telephone company is now available from third parties. This approach also has the advantage of being consistent with the "layered" nature of the Internet's architecture. Arguably, the focus of interception should be at the transport layer, not at the application layer, and the provider of transport services should be obligated only to isolate and deliver to law enforcement the data stream associated with a particular subscriber. This could be coupled with technical and legal audits to ensure that the government is only recording what it is legally authorized to intercept.

Conclusion

Congress has taken a relatively non-regulatory approach to the Internet and has refrained from applying to the Internet common carriage status and other regulatory burdens applied to telephone companies. The Internet's rapid growth and innovation attest to the wisdom of this policy. We are now in a time of transition from the narrowband, dial-up Internet of the past to the broadband Internet. The high speed Internet access available via cable modem and digital subscriber lines (DSL) is capable of carrying voice communications of high quality, as well as numerous other applications. This is precisely the wrong time to shoe-horn the Internet into the telecommunications regulatory structure.

The Internet and applications like Voice over Internet Protocol (VoIP) services are different from traditional telecommunications services, so significantly different that they have not been and should not be regulated under the traditional regulatory framework for telecommunications. For reasons that are still valid today, the Internet and Internet applications were not included in the regulatory mandates of CALEA. After an in-depth factual inquiry in the early 1990s, Congress focused on specific problems law enforcement agencies were encountering in carrying out surveillance in the PSTN. With CALEA, Congress imposed design obligations on already heavily regulated telecommunications common carriers. Congress expressly excluded the Internet from those design mandates, not only because it was committed to the non-regulatory approach, but also because it found no problems on the Internet, and because it was uncertain of how surveillance mandates would translate to the Internet.

The regulatory framework of CALEA is not suitable for the Internet and Internet applications. The FBI and the Justice Department are absolutely correct when they say that the world of communications has changed dramatically since CALEA was enacted. That is exactly why applying a 10-year-old law to this rapidly evolving technology would be a mistake. CALEA-type mandates would drive up costs, impair and delay innovation, threaten privacy, jeopardize Internet security, and force development of the latest Internet innovations offshore.

Most importantly, the centralized design mandates of CALEA are not necessary. The government itself can acquire the technology it needs to interpret Internet communications. It will have to do so in case, because there will always be custom-built services and applications outside its reach. The sooner it abandons its efforts to dictate surveillance features to industry, the sooner it can get on with the task of keeping pace with technology.

Mr. UPTON. Thank you.

At this point members will be able to ask questions for 5 minutes.

I just want to, Mr. Dempsey, go back to what you just said in terms of talking to the cable industry. The cable industry, and I think, Dr. Green, you will support, I mean, as you look at VoIP it is viewed as telecommunications, and therefore, you are subject to CALEA; is that not correct?

Mr. DEMPSEY. Yes, Mr. Chairman, that is correct. I think the cable companies early on, when they viewed the possibility of offering voice services on their networks recognized that it was going to be necessary to comply with CALEA, and that is one of the reasons that they tasked us, Cable Labs, to develop the specifications or to make that a viable technical solution.

Mr. UPTON. And to underscore or reiterate your testimony, you indicated that all of the cable companies are compliant; is that not correct?

Mr. DEMPSEY. As far as I am aware, all are compliant.

Mr. UPTON. Now, Ms. Parsky, you indicated in your statement and you said some carriers without regard to court ordered CALEA, some are and some are not complying; is that correct?

Ms. PARSKY. That is correct.

Mr. UPTON. Can you help us in terms of who is not being helpful? You did not name anyone by name. Obviously the cable companies appear to be based on Dr. Green's testimony.

Ms. PARSKY. Well, we have purposely not named people by name, because we recognize that it is important for us to try to work with the companies that are not compliant to try to get their cooperation. I can tell you that there are companies out there that when they have been served with interception orders have not had the capacity built into their networks to be able to comply with the court orders as required by CALEA.

Mr. UPTON. And how would you, Mr. Thomas and Mr. Knapp, respond to Mr. Baker's comment that it is up to you to figure it out versus the companies themselves to comply with CALEA?

Mr. KNAPP. CALEA provides that the carrier can select their method of complying with CALEA by either using a standard or whatever method satisfies law enforcement, and in many cases, these standards have not been fully developed or the carrier has argued that they needed more time to comply with the standard.

Mr. UPTON. And you have given them that time; is that not correct?

Mr. KNAPP. That is correct.

Mr. UPTON. Mr. Thomas?

Mr. THOMAS. Well, from our perspective I would say, first of all, our requirements or needs in terms of law enforcement interception were developed and are developed with an eye toward lawful interception. So what we set out as the elements out of a network are based on information that we need to make rational decisions over whether we are lawfully authorized to monitor that conversation or not. So that is where our needs develope from.

There is no doubt that CALEA did not necessarily accomplish one of its main goals, and that was to create an atmosphere in which early on in design time carriers were able to reach out and grab a ready made standard and implement it so that when the new services were offered, there was no conflict and no problems. It hasn't created that atmosphere.

Unfortunately, we do need that sort of an atmosphere, but we do have to also recognize that the law enforcement community has needs that are based upon the authorizing statutes for doing electronic surveillance, and that is what we are trying to implement, and we are trying to rely on the carriers who best know their technologies to do that.

Mr. UPTON. Ms. Parsky, do you want to comment?

Ms. PARSKY. Well, I think that Mr. Thomas has expressed this well. One thing that I think has been distorted is the fact that CALEA provides for the FCC to determine what standards need to be met in implementing CALEA. It is not the FBI, and to the extent that the industry and the private standards bodies are creating these standards, if they come to the FBI and say, "Do these standards meet your needs, your needs that are determined by, you know, your authority to intercept these communications?" and when the FBI relates that some things do and some things do not, if the industry wants to challenge that and wants to have the FCC step in and determine what is required, it is the FCC that makes that determination.

Mr. UPTON. Dr. Green, as you indicated the cable industry is 100 percent compliant, how do you respond to Mr. Baker's comment that, in fact, going back to the Wall Street Journal story that, in fact, it may drive that innovation overseas?

I mean, has that happened in the cable industry? It did not, did it?

Mr. GREEN. Well, Mr. Chairman, that has not been our experience. Obviously, these technologies are very competitive, and they are certainly competitive worldwide. So there are risks to technology development here. However, our experience has been in working with U.S. manufacturers and in working with the FBI, we were able to reach agreements which I think are in themselves innovative solutions.

I believe that the technical details contained in our specification amount to new innovative approaches that were developed as a result of trying to work together to solve problems.

Mr. UPTON. Thank you.

Mr. Wynn.

Mr. WYNN. Thank you, Mr. Chairman.

Just a couple of questions. If I heard him correctly, I believe Mr. Dempsey said that there was not an adequate factual record in terms of what the specific problems were. So I wanted to ask Mr. Thomas and also Ms. Parsky if you agreed with that statement, and if not, whether you believe it ought to be on the record or if you agree it ought to be added to the record. If you disagree, why do you disagree with that?

Ms. PARSKY. I think it is important to point out here that in terms of a factual record of where, exactly where, law enforcement is having trouble effectuating court orders and conducting intercepts is something that is extremely sensitive for law enforcement. To have to lay out a record of exactly which services are the places that criminals and terrorists should migrate to because that is where law enforcement is struggling is something that requires the appropriate setting to lay that out.

I think the other thing that is important to recognize is that once there is a lengthy record of all the times that law enforcement has been unable to effectuate court orders for interception, that is too late. That is when all of those criminals have already had the opportunity to do harm to our society.

Mr. WYNN. Let me interject then. If I am understanding you correctly, you are basically saying that this material would be classified. That record that Mr. Dempsey says ought to be established would be classified material, and perhaps not in this forum, but in an appropriate governmental forum you would be willing to provide that so that Members of Congress could evaluate the record and the extent of the problems. Is that a fair assessment of what you are saying?

Ms. PARSKY. I think that a fair statement is that a great deal of that record would be classified and we would not be able to share that in this forum.

Mr. WYNN. Is there a forum within Congress that you believe would be appropriate?

Ms. PARSKY. In a classified forum.

Mr. WYNN. A classified forum. Would you agree with that, sir?

Mr. THOMAS. Yes, I would agree with that.

I would also just add it is important for us to establish a factual record, but the factual record is not only made up of specific cases where we have encountered problems, but also it is based upon a knowledge of electronic surveillance over many, many years and our understanding of where criminals are likely to move and terrorists are likely to move or have already moved.

So I would support Ms. Parsky's statement that a part of this is also predictive.

Mr. WYNN. Well, I think those two parts are classified and the predictive part ought to be presented to Members of Congress at an appropriate forum so that we can then, as Mr. Dempsey said, evaluate exactly what the problems are.

Mr. Baker, you were suggesting that a system that would require prior FBI approval would be so burdensome as to drive innovation abroad with respect to market testing if I understood you correctly.

What exactly is the nature of the burden? Is it the time to be consumed in getting the approval? Is it the demands of the standards that would be required prior to a market test in this country, particularly in light of the fact that Dr. Green seems to suggest that you would not have that effect using an FBI approval system?

Mr. BAKER. I think the biggest risks come in the Internet context where today if you have a good idea, you just put it out on the Web and see if people come and flock to it. Hotmail started as a guy with a good idea. He said, "Gee, why don't we offer free E-mail accounts in exchange for people looking at ads?" and that became a $400 million business in about 9 months.

He did not have to get anybody's permission. He just started the business. If, instead, he has to stop, say it is a voice version of Hotmail, he has to stop and say, "Well, do you think the FBI would view that as covered by CALEA? What would have to give them exactly if they thought it was covered by CALEA? How would I design that into my system?" you are adding months of design time, and then he has to hire a lobbyist. He has to come to Washington and hire somebody to deal with the FCC. All of that is a burden on his ability just to start a business.

Mr. WYNN. Let me interject because it seems like it is somewhat speculative. Could he just go into the FBI and say, "This is my idea. It is patented or whatever, but what do you think? What do you need?"

I guess I am wondering whether this is actually a month long problem, particularly if it is only a question of whether he can market test it anyway, whether or not that cannot be accomplished in a shorter, more reasonable period of time.

Mr. BAKER. That is a very good question. I think the answer is the FBI is not used to being a regulator. That is not what they do. They are not used to taking risks. I think what would go through the mind of the FBI office that was asked that question is, well, if I say yes and it turns out badly, am I going to get blamed. The answer is yes. If i say no and it turns out badly, am I going to get blamed? And the answer is, no, no one will care because they will not hear about it.

So all of the incentives are to say no. We need more information. We cannot approve this yet. We want to see more specs., more design, more features.

You know, I think John L. Lewis, when he led a wartime coal strike, was asked what do the miners want and he said, "The miners want more." And in my experience in dealing with the FBI, that is often their position. They just want more. they want as much as you can do.

Mr. WYNN. Could I get just a response from the FBI as to whether that is an accurate interpretation of how this process might work?

Mr. THOMAS. Well, I would not agree with all of it the way it was said, but I would say that the CALEA statute, the way it was written, basically puts everybody on equal footing if CALEA applies to them. So, it is essentially, you are in it or you are out of it, unless you can find a statutory exemption.

So it really does not put the flexibility necessarily within the FBI's hands or the FCC's hands to say, you know, you are out but you are in because, in part, it laid a level playing field out for those types of carriers.

Some other nations we have seen have applied laws in a different way that either had what you might call opt in or opt out policies where you have exactly that procedure. Unless a government agency tells you you must comply, you don't have to comply, or once a government agency tells you or you have to consult and somehow get a buyout, that is not the way CALEA as we understand it was implemented.

Mr. WYNN. Thank you, Mr. Chairman.

Mr. UPTON. Mr. Walden.

Mr. WALDEN. Thank you, Mr. Chairman.

Mr. Thomas, how do you respond to the 2004 DOJ Inspector General's report finding that the FBI, and I quote, was unable to demonstrate the extent to which lawful surveillance has been adversely impacted by the lack of CALEA implementation?

Mr. THOMAS. Well, I think my response to that is that early on with CALEA we recognized that there would not be enough money in the fund set up for it to cover the Internet 100 percent. We knew that before it was appropriated.

We also knew that there had to be a process, and this was statutorily set out for deciding which technologies that were already in place we would pay for and which one we would not.

So we had to make decisions, and that is the one procedural place where we had decisionmaking on who was in and who was out. Fairly early on, we made the decision to focus more intensely on areas where we would be completely shut out of any capability very rapidly if technology moved the way we expected.

It turns out technology did move the way we expected, and it was the wireless world. We focused much more energy and much more cost from our perspective on making sure those capabilities were in place, and I quoted a number that roughly we can calculate from the wiretap report about 70 percent of authorizations were for CALEA covered, CALEA capable services, and that is a calculated number based on how many were wireless and how much wireless we have covered.

But the point is, our goal was to apply the money we had and the level of effort that we had to try to get the best benefit, and I think that is what we have done.

Mr. WALDEN. Well, perhaps a question for both you and Ms. Parsky. How do you respond to Mr. Baker's contention that, and I believe I am quoting correctly, despite the crisis atmosphere fostered by the government, the Justice Department law enforcement have never once used the enforcement powers that CALEA gives them?

Did you just say 70 percent of the wiretaps are under CALEA?

Ms. PARSKY. Well, if I may, I think those are referring to two different things.

Mr. WALDEN. All right.

Ms. PARSKY. I believe it is Mr. Dempsey who quotes the Inspector General's report with respect to software, CALEA compliant software only being activated on 10 to 20 percent of wire line technology, and Mr. Thomas was pointing out that that is a misleading quotation because where we focused our efforts was where we thought the technology was migrating, which was to the wireless technology, and for wireless, I think these are rough estimates, that about 90 percent of the FBI's intercepts are over wireless phones, and 80 percent of those 90 percent—or the other way around: 80 percent of the wireless phones and 90 percent of those wireless intercepts are CALEA compliant.

Mr. Baker's reference to enforcement, I think, is to the extent that the Justice Department would go to court to enforce a company's failure to comply with CALEA. And in that regard, what we have done is we have tried to go to the companies and work with them; because we realize that in the long run, rather than going through a lengthy and expensive process for them, to the extent that we can work to show them that, in fact, their technology, their services fall under CALEA, the importance to national security and public safety that they be able to carry out the intercept orders, that we want to work with them to get them to do that.

And it has been a slow process, and our filing the recent petition with the FCC was part of our having to take a step that is more forceful than just going and trying to work with the companies.

Mr. WALDEN. All right. I represent a very, very rural district. It is sparsely populated, a lot of small exchange carriers, and one of them just installed a new switch, and they were looking, you know, at CALEA and some of the requirements for voice over Internet protocol and said, "We have never had a wiretap in 30 years in this little town and now you are going to have this new requirement in on top of us." I would be curious from the panel. How do you respond to that? What do I go back and tell these folks out in very rural Piece of Heaven?

Mr. KNAPP. First of all, there are provisions in CALEA for requesting extensions of time, and beyond that, if they can make an argument for various reasons that the costs are not warranted, they have that opportunity to do that under CALEA.

One of the things we tried to do in our rulemaking proceeding is identify not only a standards approach to comply, but also other solutions, such as use of a third party that might make sense for people who only get an occasional wiretap to engage the services of a company that could help them meet the requirements of the court order.

Mr. WALDEN. And, Mr. Baker, do you have anything to add?

Mr. BAKER. Yes, if I could just add, I do think though that there is a problem here that shows why the FBI is just not suited to this regulatory role. It was too hard for the FBI to say, "Yeah, we know. We will probably never need to do a wiretap in your district. So we will give you a pass."

Because from the point of view of the person who is asked to do that if that turns out to be the wrong decision they will be blamed, but if they make a decision that is too regulatory, it does not cost them anything, and I think you see that bias over and over again in the way the FBI has approached the bill.

They do not ever want to be wrong in a way that might cost them one wiretap, and so they tend to over regulate even in circumstances where it just doesn't make economic sense.

Mr. WALDEN. Mr. Knapp, maybe I can go back to you on the technology itself then when you talk about a third party provider. Is this a patch that they can come in and put in and take out?

Mr. KNAPP. In some cases it may involve implementing software in existing equipment that separates out the subject's data stream from everybody else's in part to protect privacy, then if necessary to convert the information so that you can pull out the call identifying information, and then prepare it in a format that is usable on law enforcement's equipment.

Mr. WALDEN. But I guess the point is is it something they can come in and insert and then remove? Could you hire somebody to come do that with your equipment?

Mr. KNAPP. That would be an option.

Mr. WALDEN. Okay. My colleague and I were talking. Given the international scope of all this, how do we regulate the Internet here and not have somebody else dump software out there that everybody could download and use and get around CALEA?

Mr. DEMPSEY. That's an important question, again, about what price are we paying and then are we really getting the result that we desire. The FBI is always going to have to have the ability in house to try to deal with customized services peer to peer in the NPRM, and the Commission recognizes that peer to peer is not covered, that there are a range of other services not covered. Traditional dial-up Internet access would not be covered.

And in all of those cases the FBI still has the legal authority to wiretap, and they have the technical capability to access that stream because the issue really here, Congressman, is not about content. All of the service providers are fully capable of providing the content. It has never been a dispute. The question is opening up those packets, breaking them apart, analyzing them, figuring out when you do an ordinary wiretap on a regular old fashioned phone line whether it is a fax or a phone conversation that is coming over the line.

The FBI has always had that responsibility to bring it in and analyze it. A little bit of the debate here is trying to push some of that responsibility onto companies. But at the end of the day, the law enforcement agencies are still going to need that in-house capability for all of the other kinds of unique things out there and for all of these other services.

I think that is a better place to look for a solution generally rather than taking this kind of very regulatory, very specific mandate forcing it on a few carriers when you're still going to need that in-house capability to deal with customized offshore, et cetera.

Mr. WALDEN. All right. Thank you.

Mr. UPTON. Mr. Buyer.

Mr. BUYER. I was here in Congress when we did CALEA, and I remember the balance tests that we were all struggling with. I have tried to be a good listener to two other opinions, Mr. Dempsey and Mr. Baker, and at the same time I am rather curious why the dispute resolutions that Congress had set out are not working.

Why is it necessary for you to make this petition to the FCC? Because that is highlighting that perhaps the framework that we laid out back then isn't working well and why it isn't working well.

So, Ms. Parsky, when you testified just a second ago, you said it is difficult in working with the companies. You have to articulate that better for the justification. So could you give us some idea? What do you mean it is just too difficult to work with the companies?

Ms. PARSKY. Well, I am not sure exactly which part of my responses you are referring to in terms of "too difficult to work with the companies," but I believe what I was trying to portray was the fact that we are trying to work with the companies, and some companies are more difficult to work with than others.

And what led to our filing the petition is the fact that we are 10 years out from the passage of CALEA, and technology has moved; and it is our argument that the direction it has moved has made it clear to us that these new technologies fall within the mandates of CALEA.

I think one of the things, I mean, there are several things that were——

Mr. BUYER. But industry recognizes that, too, that they have to be compliant.

Ms. PARSKY. Some do and some do not. Some have just deployed their technologies, presented them to the public, and then decided that they would argue it out, either wait for us to confront them on it and argue it out through the FCC, which would take a long time and then through the court system, and in the meantime they would be able to reap the benefits from having the technology out there. Where for us what that means is there is something out there that does not have the capabilities built in to comply with court orders.

Mr. BUYER. So what did we do wrong with CALEA? What should Congress have done right?

Ms. PARSKY. Well, we are in the process of really trying to evaluate what it is that we would need as an updated version of CALEA going forward, because technology has changed so much.

But I think one important thing has caused confusion in CALEA, and I think that is evidenced in Mr. Dempsey's statement when he says that CALEA was not intended for the Internet.

And I think that there is an exemption in CALEA for information services. It is not an exemption for the Internet. It is not an exemption for Internet services. It is for information services as information services were defined. But those were information services that were storing E-mail, that were static Web sites, but not the transmission of information over the Internet.

The transmission of information even back 10 years ago was recognized as something that was not included in the information services. At that time it was through dial-ups, so that the transmission would have been over wire lines, over telephone lines.

Now that transmission is over cable lines. So I think it is the confusion over the sort of parsing out of information services and the changes in technology and how the Internet is being used today that has caused a lot of the confusion.

Mr. BUYER. Well, I will concur with you that Congress did not give you any wiretaps authority, but what we really intended was for you to have access, and you know, we do not intend to be stifling the technologies out there. I mean, we look at at the Telecommunications Act, and it has been a success by getting regulation out of the way and letting innovation just really take over.

But you are outpacing law enforcement. I mean, that is the reality. That is what our present struggle is.

Mr. Baker and Dempsey were pretty tough on you. Do you have any comments that you would like to make in response to their testimony?

Mr. KNAPP. A couple of points because I think there are some misunderstandings of what the Commission actually proposed.

We rejected the notion of a pre-approval process for technologies. We absolutely do not want whatever steps we are taking here to constrain technology. What we are trying to do is be very specific about what the capability requirements are so that everybody understands going in what they are required to do.

Leave it to industry to develop the standards, but have certainty for the carriers so they can be assured that if they comply with the standard that it won't be challenged.

So on the point that Mr. Dempsey raised, as we look back at the 1994 act, we see an irreconcilable tension where, on the one hand, its intention was to maintain law enforcement's capability to perform wiretaps and to exclude information services. It was not envisioned at that time that the Internet was going to become a means for making telephone calls, but the act provided for looking at advanced services as they unfold and for the Commission to make a public interest finding in those cases as to whether those services should be covered, and that is the purpose of our rulemaking.

Mr. BUYER. So that I am not left with a bad impression, Mr. Knapp and Mr. Thomas and Ms. Parsky, maybe you could help me with this. Is it sort of the presumption by industry now to go ahead and deploy and litigate, or I have got my technology; let's try and work it out with law enforcement?

Where do you think it stands right now within the community?

Mr. THOMAS. I think from my experience it is the latter, and that is, that for the most part, industry is moving forward with innovation. They already in many cases have people thinking about CALEA capabilities. Some companies just do not believe it applies to them. Others do, but for the most part it is not simply we are going to roll this out and see what the Justice Department will do.

I think it is an idea of getting it out there, moving forward, and then coming behind that and trying to work with standard bodies, work with law enforcement. I do not think there is that much disingenuous effort out there. I think there is real effort.

Mr. BUYER. You know, a lot of times here in Congress we make laws, but because of exceptions, and that is just what I am trying to figure out if there is a much larger problem or if we are trying to create something because of a smaller community giving you a hard time.

Can you help me out here?

Mr. THOMAS. Well, I do not think that CALEA was a badly crafted law for the time. It did not have as much agility in it, I think, as we are going to need going forward in the Internet world. It is a different environment, very different, and I think there are going to have to be tweaks to create agility where it does not exist.

Mr. BUYER. We thought the whole future back then was voice.

Mr. THOMAS. Right, and a lot of people still——

Mr. BUYER. And everybody at that table if you were willing to admit to yourselves, we did, too, and we got it wrong. It is all about data. That is the power.

Mr. THOMAS. And a lot of people still to this day focus on voice when they talk about CALEA, but CALEA really did not focus on voice. It focused on wire and electronic communications, which was very broad, but it did not quite have the agility we need, and I do think going forward it is utterly impossible to do what we need to do without cooperation. It is utterly impossible.

And if we do not have a legal structure that encourages or somehow causes or urges cooperation from industry with law enforcement and vice versa, it will not work.

Mr. BUYER. All right. Thank you, Mr. Thomas.

Thank you, Mr. Chairman.

Mr. UPTON. Thank you.

Mr. Shimkus.

Mr. SHIMKUS. Thank you, Mr. Chairman. I will be brief.

I have a district where I have quite a large rural size, too, and so I echo some of the comments from Mr. Walden, and I would just make two points of observation. One is that I do believe in the market. So if a small, rural telephone company, family owned, which we still have them, or co-op, you know, have challenges, and I understand the law allows provisions to delay or that, you would think the market would also fill the void with specialists who come in.

If there was one every 30 years that if I was a smart guy, I would be figuring out how to be that repairman or that specialist that would come in and help comply with a legal court order.

But the second thing is understanding the threat. I would think that in the digital age crooks and criminals would go to the areas where there is less risk of being able to be tapped technologically. So a rural area might be a great place to set up your scam to take money from one Account A and transfer it electronically to Account B with the time lag of trying to put in the provisions the legal authorization, you know, to have access to those communication lines and then actually technologically have them in place to go after the crooks or criminals or in the world that we live in, terrorists.

So this is a good hearing. I do concur with my colleagues that technology is moving quicker than our ability, and we will have to be in this together.

Mr. Baker, I think.

Mr. BAKER. Yes. I thought I would add to a response to Mr. Walden's question and yours about the rural carriers. I will tell you what I would have said to them under CALEA, as I understood it before the FCC ruled. I would have said, and in fact, I have said, if the FBI comes to you and says, "We have to do a wiretap. We have to hear what this guy is saying. Can you do that?" the answer is almost always, "Yes, we can do that."

We can find a way to plug the system so that we can give them the one wiretap they need every 30 years. What we cannot do is give them all of these party-hold, party-join, party-drop, crazy stuff that is in the CALEA standards that the FCC and the FBI have approved. We just do not know how to get that information.

And I would have said, well, if in 30 years you get one of these and you can provide them with the content, and they do not like that, which is unlikely, then their recourse is to sue you, and you will be able to defend on the ground that it is not reasonably achievable. You could not be expected to spend that kind of money for one wiretap after 30 years, and that there were other places that they could get the information that they needed.

Those are all of the defenses that the statute creates for them, and it forces the FBI to actually justify what it is doing. Under the regulatory proposal that the FCC is putting forward and the FBI and the Justice Department are supporting, I would have to say to them, well, if you do not have that and the FBI walks in with or without a wiretap, they can ask for fines to be imposed on you tomorrow. So you have to go out and spend the money right now.

And that is the difference between what CALEA intended and the regulatory proposals that we are seeing out of the agencies today.

Mr. SHIMKUS. But the big concern, I think, those who were here when CALEA was initially passed was the ability of law enforcement to have access when there is presumed bad guys using the systems, and the threat, especially the international terrorist threat and al-Qaeda operatives. I don't know if we want to wait. That is the dilemma we are in.

The law enforcement, the FBI and all of these folks who are trying to protect us, you are right. I mean, they do not want to be on the hook for being wrong one time, and do you know what? We do not either. We want them to have it right the first time to hopefully preclude what has gone on in the past.

Mr. Chairman, that is all I have. I yield back.

Mr. UPTON. Mr. Cox.

Mr. COX. Thank you, Mr. Chairman.

I wonder if I could drill down a little bit on the software side of this. I wonder if any of our witnesses feels comfortable telling us this morning what would be the process for a software developer to bring its product to market if it includes the voice capability.

Are they going to have to go to the FCC, to the FBI, to the Department of Justice? What do they do?

Mr. KNAPP. It would be as simple as complying with an industry standard. They would not have to come to the FCC for approval to do that. So, for example——

Mr. COX. Well, wait a second. An industry standard can only exist for things that have been agreed upon in advance. What happens if I am an innovator?

Mr. KNAPP. The industry standard can be generic in terms of providing information about where is the starting point of the communication and the endpoint, without getting down into the applications itself.

Mr. COX. And is this something that has been included in the Justice Department petition? Have you specified exactly how that would work?

Ms. PARSKY. What we made clear in our petition is that CALEA does not give law enforcement, the FBI, the Department of Justice the ability, nor should it, to dictate what the standards are. Those who are in the best position to innovatively devise the best ways to meet law enforcement needs are industry themselves. So that to the extent you——

Mr. COX. So right now we are in a position where we do not have standards. We are hoping we can develop them. The standards themselves would have to encompass known technologies. They could not really credibly encompass things that had not yet been invented.

And so anything new does what? Something that is not definitionally covered by the agreed upon standard that encompasses all existing technology?

Ms. PARSKY. Well, to the extent that there are new technologies and companies had questions and want to work with the FBI for their assistance, we are more than happy to do that.

Mr. COX. I want to work with the FBI. What do I do? Where do I go?

Ms. PARSKY. Mark, do you want to?

Mr. COX. Does the FBI have a private sector innovation office where I can go and meet with people who are happy to help me get my product to market?

Mr. THOMAS. Well, I mean, there are two different issues here. One, I think, is software developers who develop applications. We generally do not see those people with CALEA requirements on them. It is normally the people who offer services of some kind to the public, as CALEA states, as common carriers for hire essentially.

For those types of organizations, we have——

Mr. COX. Well, hold on a second.

Mr. THOMAS. Yes.

Mr. COX. The Justice Department petition to the FCC asks that the FCC rule that broadband Internet access be covered by CALEA. That is at least according to your testimony, Ms. Parsky. And so if I am now the provider of broadband Internet access, I have got to begin worrying about all of the software that is going to be used on my system so that I can comply with my legal obligation, and it is not distinguishing at all, and I do not know how it could, between the things that are agreed upon under the standards and the things that are brand new.

So I have got to make sure that the new things which I do not have any legal comfort on are bumped from my system and go through some queue, and then I am going to send them to the FBI or wherever, and then I want to know who is going to be in charge of that.

Mr. THOMAS. Well, I mean, I think the point there is for the broadband provider. What we are saying is the obligation on him is clear at least with regard to whether or not he has to isolate and provide to law enforcement as CALEA requires call content.

With regard to the call identifying information, which is really what you more seem to be pointing at and who understands how to reach you and pick out that particular information, I do not think that is an issue that has actually matured yet. I do not think anyone has a complete vision of how that will be carried out.

I think it is probably likely that there are some services in which the provider themselves, the access provider, say, a cable company, is in the best position to do that work. There are other services where they may not be in such a good position to do that work.

There will be some applications for which no one really will be in a good position to do that work, and ad hoc solutions will continue to have to operate in those areas. I think that is a bit of a balancing act that we will expect to have to——

Mr. COX. Mr. Chairman, here is what I am concerned about. Congress and this committee, we in this committee, expressly excluded information services from CALEA because we did not want to have this collar on innovation in advance. As information services and telecommunications services morph into one another, we are going to have some tough calls to make, but we surely did not make them in the statute, and we also had a paradigm in mind in which the digital innovation that we were thinking of as information services was excluded expressly in the statute.

If we are now going to finesse all of this in a regulation without amending the law, I have not any comfort that we will not—particularly if the petition from the Justice Department asks that the gateway for everybody's Internet service, the broadband access, you know, be covered by CALEA—I will not have any comfort that we will not be basically saying we are going to treat all information services, all your E-mail, all data transmissions, and so on the same way that we treat voice-grade telephony for purposes of wiretap.

When Alexander Graham Bell went out and strung some wire and began to exploit his invention of the telephone, there was not any wiretap scheme in place. The order was innovation first, regulation second.

What we are talking about now is a fundamental shift. We are going to go to regulation first and innovation second, and virtually all of these innovators are going to have to queue up single file before a government agency in Washington before they can come to market, and that really does give us Mr. Baker's problem, which is that we push this stuff overseas.

And incidentally, some of the people who want to create problems for us might well be developing software into which they imbed their own listening algorithms and then dump it onto our market in this way. People pick that up because it does not bear the stamp of, you know, the U.S. Government wiretap approved.

I think we ought to be very cautious about entering this area, and certainly, Mr. Chairman, we ought to be very cautious about doing it with regulation.

Mr. Dempsey, sorry.

Mr. DEMPSEY. Mr. Cox, I think in your dialog with Mr. Marcus you have identified something very important here, which is the FBI, Department of Justice at this point cannot even identify who will bear what responsibility. So they are asking to regulate all broadband access and all managed VoIP services without even knowing who will have what responsibilities and what those responsibilities will look like.

So we are putting the power of CALEA and the whole pressure of CALEA on top of entities that we have not even really identified yet, and we do not know what their responsibilities will look like.

I think that one of the steps here should be, in addition to identifying the problems, the second half of it is what will the solutions look like, and if you regulate first and bring either by regulation, as the FCC is proposing, or by legislation, if you bring Internet services into the scope of CALEA, you had better know in advance what those obligations are going to look like and how they are going to translate rather than leaving that to the FBI and the FCC to work out.

Mr. COX. Well, I think my time has expired. I would just say, Mr. Chairman, that innovation is not necessarily the enemy. Technological innovation, which America leads in so many respects, is the best advantage we have going for us in law enforcement, and so we do not want to stifle it needlessly.

I yield back.

Mr. UPTON. Thank you.

Mr. Stearns.

Mr. STEARNS. Thank you, Mr. Chairman, and thank you for having this hearing.

Mr. Knapp, the term "telecommunications carrier" appears in both the CALEA and the Communications Act, with CALEA's definition being much more broad and enveloping. If the FCC determines a certain company is a telecom carrier under CALEA, does that sentence the same company to the Title II designation under the Communication Act?

If not, what conflicts would separate designations under these statutes, so to speak?

Mr. KNAPP. The short answer is no. A determination relative to CALEA would not necessarily have any implication relative to the Communications Act. Our decisions and the proposals that we made are based on the unique provisions of CALEA, which provide that the Commission may in the public interest include services that are a substitute for public switched telephone service.

So we do not believe that the CALEA obligation carries with it any implication that the Commission may apply the legacy regulations that we have in the past under the Act.

Mr. STEARNS. How does the NRPM address peer-to-peer services that may offer VoIP be covered?

Mr. KNAPP. It suggests that this is an issue that we need to resolve. We propose to include managed services. These are services——

Mr. STEARNS. What are they?

Mr. KNAPP. They are services where there is a party in control of setting up the call and potentially controlling the quality of the call. We did not include peer-to-peer service which at the extreme end could be simply two people putting software on their PCs.

Mr. STEARNS. One of the goals that Mr. Boucher and I sought in the bill we developed and dropped was to redefine these new IP based services so that we might be able to address some of these same questions, and so we proposed a new definition in the context of a broad rewrite of the Telecom Act.

Do you think that would be beneficial?

And I would like to ask all of the witnesses if they think maybe a redefinition in the Telecom Act of 1996 to include this broader definition. I do not mind it if you do not agree with me. So go ahead.

Mr. KNAPP. I have no position on that.

Mr. STEARNS. No position? Even a personal?

Mr. KNAPP. No, not even a personal. Our focus here was solely on CALEA and the provisions of CALEA.

Mr. STEARNS. Okay. Why don't we just start? Would anyone else have a comment on this?

The idea is to address some of the questions we might redefine this new IP based service. Yes.

Ms. PARSKY. Well, in terms of a broader definition of the services, yes, the Department of Justice does not have a formal position, but we would be happy to sit down and meet with you and discuss the definition with you.

I think that the one concern we have with the bill that has been proposed is an exclusion for regulations such as CALEA and that, and I think we have addressed fully in our statements here.

Mr. STEARNS. I think we did that because of jurisdiction. So it is nothing intentional.

Anyone else? Yes, Mr. Baker.

Mr. BAKER. Yes, TIA's view is that VoIP is an information service and ought to remain so. As for CALEA, we would really encourage the committee if there is something that needs to be fixed, instead of kind of schmoozing it fixed through the FCC, the Justice Department ought to come up here and propose regulations or legislation that would fix the problems that they see.

And we would support having an examination of new technologies to see what more needs to be done. It is just that we do not think that trying to slip it in through the back door through the FCC regulations is the way to do it.

Mr. STEARNS. Good point. Okay. Mr. Dempsey or Mr. Green, anything?

Mr. DEMPSEY. Well, Congressman, I would just say that we are winding up caught in a straightjacket of telecom services, on the one hand, information services, on the other hand. In all likelihood, those definitions do have to be reexamined, and along with them, the regulatory burdens that they carry or the public policy interests that are served both on the economic side and on the social policy side, and I think there the question for this committee is going to have to be what problems are there that need regulatory action in the first place.

For most of the history of the telecommunications system, there was no problem with carrying out wiretaps. There was no regulation at all. It was an almost incidental byproduct of the design of the networks.

In 1994, Congress concluded that with respect to traditional telephone services, there were identified problems. And now the question is with respect to new services, however you categorize them from a regulatory, whatever word you use to describe them, what is the problem, if any, and it has still not been shown that there is a problem, and then what does the solution look like.

Mr. STEARNS. Mr. Chairman, my time has expired. I ask unanimous consent to ask one more question.

Thank you for your kindness.